Texas Cacti

NUMBER FORTY-TWO:
W. L. Moody Jr. Natural History Series

Texas Cacti

Brian Loflin & Shirley Loflin

Texas A&M University Press   College Station

Copyright © 2009
by Brian K. Loflin
& Shirley A. Loflin
Manufactured in China by
Everbest Printing Co.
through Four Colour Print Group
All rights reserved
First edition

This paper meets the requirements
of ANSI/NISO Z39.48-1992
(Permanence of Paper).
Binding materials have been
chosen for durability.

LIBRARY OF CONGRESS
CATALOGING-IN-PUBLICATION DATA

Loflin, Brian.
   Texas cacti / Brian Loflin and Shirley Loflin. — 1st ed.
     p. cm. — (W. L. Moody Jr., natural history series ; no. 42)
   (TAMU nature guide)
   Includes bibliographical references and index.
   ISBN-13: 978-1-60344-108-7 (flexbound with flaps : alk. paper)
   ISBN-10: 1-60344-108-5 (flexbound with flaps : alk. paper)
   1. Cactus—Texas—Identification. 2. Cactus—Texas—Pictorial works. I. Loflin, Shirley. II. Title. III. Series. IV. Series: TAMU nature guides
QK495.C11L58   2009
583'.5609764—dc22
2008042581

p. i Graham's fishhook cactus (*Mammillaria grahamii* var. *grahamii*)
p. ii-iii Sea urchin cactus (*Coryphantha echinus*)
p. iv Strawberry cactus (*Echinocereus stramineus*)
p. v Brown-spined prickly pear (*Opuntia phaeacantha*)
p. vi Strawberry cactus (*Echinocereus stramineus*)
p. viii Brown-spined prickly pear (*Opuntia phaeacantha*)
p. x Typical Big Bend hillside vegetation includes the commonly occurring strawberry cactus (*Echinocereus stramineus*) and Engelmann's prickly pear (*Opuntia engelmannii* var. *engelmannii*). Also typical within the Chihuahuan desert habitat are indicator plants, creosote bush (*Larrea tridentata*) and lechuguilla (*Agave lechuguilla*) as well as ocotillo (*Fouquieria splendens*).
p. xii El Capitan, 2447 m (8,085 ft), greets travelers as it dominates the landscape of far west Texas.
p. xv Brown-spined prickly pear (*Opuntia phaeacantha*)
p. xvi Cow-tongue prickly pear (*Opuntia engelmannii* var. *linguaformis*)

This book is lovingly dedicated to Harvey and LaMar Welton, friends from Riverside, California. Owners of a small, yet prolific cactus nursery, they have been most generous with their wealth of knowledge of cacti and succulents.

Almost instinctively during our first meeting while acquiring specimens for our own garden, we knew this couple would become dear friends. We are most grateful for their unselfish and caring hearts and their love of all nature and photography. Our wonderful friendship has blossomed and thrived across the miles.

Contents

Preface	ix
Acknowledgments	xi
Introduction	xiii
Cactus Country	1
Habitat	
Vegetational Areas of Texas	
Cactus Anatomy	15
What Is a Cactus?	
Features of Cacti	
Cactus Critters	29
How to Use This Book	35
Cactus Stem Shapes	
Blooming Time	
Conservation Status in Texas	
Cactus Terminology	
Cactus Genera of Texas	43
Genera of Represented Cacti	
Species Accounts	51
Map of Texas Counties	
Appendix	259
Species Synonyms	
Bibliography	271
Glossary	281
Index	285

Preface

Many folks we visited with during this project smiled enthusiastically about the prospect of a new field guide to the cacti of Texas. Many believed that a new, concise, and colorful photographic guide was long overdue.

We produced this identification guidebook of Texas cacti as a useful tool for amateur and professional naturalists, nursery owners, hobbyists, educators, and many others interested in the natural science of Texas. We include in this easy-to-use field guide more than one hundred species of cacti found in the state and surrounding regions.

First, we divide the species by stem characteristics to provide a handy method of visual identification. We then describe significant features of each cactus along with the identification photos. The text includes references to the cactus's visual characteristics, supporting soil type, habitat, species range, and flowering season. This book, a resource combining many features of cacti in one handy text, does not have the hard-to-use identification keys of the academic botanist. It is useful and easily read, written in a language suited for many readers.

This book is designed to be enjoyed through the use of large and stunningly detailed close-up color photographs of each species presented. These close-up images of the plant stem and spine characteristics are most useful in plant identification. These, combined with color habitat images, illustrate the cacti, unlike many other reference books. Special techniques and new technology in photography and quality color reproduction provide an excellent perspective of the cacti in a way that enables easy species field identification by the cactus enthusiast.

Comments by noted botanists, cactus enthusiasts, and educators give the project praise. We hope it will become a well-used and most enjoyed publication and a worthy resource for those who care about our natural Texas environment.

Acknowledgments

A project of this scope cannot be completed alone. A vast wealth of information, plant material, inspiration, and just plain old-fashioned perspiration in the Texas sun, combined with a lot of moral support, has become the underpinning of the successful completion of this work, for which we are deeply grateful.

These caring and unselfish supporters include Cathryn Hoyt, executive director, and Marc Goff of the Chihuahuan Desert Research Institute; Patty Manning of Sul Ross University; Monique Reed of the Biology Department Herbarium at Texas A&M University; and Jim Weedin of Aurora Community College.

Thanks go to Betty Alex and Joe Sirotnak of Big Bend National Park for their guidance to unusual species within the park and for their support during our photographic trips into the park. Special kudos go to Lisa Williams of the Nature Conservancy of Texas for her perseverance on a blazing hot afternoon while capturing images of special species in the Rio Grande Valley.

We thank David Riskind, Dana Parks, and Jackie Poole of Texas Parks and Wildlife for their support and encouragement. We also thank the managers, staff, and enforcement officers of the many Texas state parks and natural areas where much of the body of this work was completed.

Our special thanks go to Jim Mauseth, a cactus specialist and professor of integrative biology at the University of Texas at Austin for his early endorsement of the project as well as for his continuous mentoring, support, and confirmation of the taxonomic effort herein.

And once again, we extend our gratitude to our friend and cheerleader, Shannon Davies, editor for natural science at Texas A&M University Press, for her belief in the two of us and the value of this project.

And to our families and dear friends that were an amazing inspiration and yet so often neglected during this time, we give our deepest thanks.

—Brian and Shirley Loflin

Introduction

Texas is big. Just ask any native. Texas is 268,581 square miles of most diverse habitat. It ranges from sea level along the Gulf Coast to 8,749 feet in elevation at Guadalupe Peak on the New Mexico border. The state is about 775 miles wide and 790 miles long at the most distant point. A drive from Orange on the Louisiana border along Interstate 10 to El Paso on the west will cover some 859 miles of diverse habitat, from forest to desert.

Within this vast state are areas of vegetation ranging from Gulf marshes, grasslands, forests, woodlands, and arid lands of thorn scrub and desert. This habitat diversity gives rise to plant diversity just as expansive. More than 4,800 species of vascular plants with over 1,300 genera are found in Texas and, not surprisingly, nearly 150 species and varieties are cacti.

Because of the great biodiversity of Texas, it has become well known botanically. Following the reports of early explorations and geographic surveys, many of the world's great botanists have worked here, collecting many species new to the science of their times.

From the age of discovery of species of cacti in the United States, from 1845 to 1883 onward, many cactus botanists have collected here and added to the growing list of cacti species, including George Engelmann of St. Louis and his associate, Ferdinand Lindheimer of New Braunfels. Other lasting names include Jean Louis Berlandier, A. Wislizenus, and Josiah Gregg. Indeed, many of our species today are named after these pioneers of botany.

Organization of cactus information was begun in earnest in the early 1900s by Nathaniel Britton and Joseph Rose, who produced a most extensive work. They were followed by Lyman Benson, Edward Anderson, and others. While this body of work was extensive, its size created somewhat confusing and contradictory taxonomical relationships. Some publications of the past two decades have created new taxonomic arrangements only to become obsolete just a few years later.

Today, as we unravel these taxonomic challenges, we are aided by the science of chromosome counts and DNA studies. This process has provided a more concise understanding of the relationships in many cactus species and has indeed rewritten extensively the taxonomy of cacti. This work, however, may not be complete.

For the purpose of this book, the authors have been guided by the impressive recent work of Michael Powell and James Weedin in the Texas Trans-Pecos and that of the Flora of North America Editorial Committee. However, the recently published *New Cactus Lexicon* is a comprehensive publication guided by the International Cactaceae Systematics Group, led by recognized cactus authority David Hunt. The taxonomic data contained herein follows the direction of this new work.

As travelers drive the highways of Texas, they frequently see the cacti of the state. They will describe the large plants with flat, pancakelike stems and lots of thorns. While highly visible at seventy miles an hour, the opuntias, or prickly pears, are widespread. However, these large plants are just the tip of the cactus community in Texas. Far more species are found if the enthusiast were to stop and more carefully examine the habitat.

Texas provides habitat for the most diverse cactus population found in any state in the United States. Many are best known by such common names as blind pear, cow-tongue cactus, night-blooming cereus, Texas rainbow, tree cactus, early bloomer, and horse-crippler, so-called because its rigid spines are dangerous to the hooves of horses and cattle. Numerous other varieties are commonly called strawberry cactus, pincushion, and jumping chollas.

In Texas and the Southwest, various species of cacti are edible and cultivated as food, notably prickly pears. The tunas, or fruit, of the prickly pear are used in making salads, wines, and jelly. The pads, or nopalitos, with their spines singed off, form a staple in Tex-Mex food. Other cacti are used to make food colorings, medicines, and candy. In some areas cacti are even used as a food for cattle.

Cacti are used for landscaping and for commercial and private botanical collections. In the neotropics some species are used as living fences, and wood from columnar cacti is used as fuel in some desert regions. Barrel cacti are known from western history as a source of water in emergencies. Peyote, *Lophophora williamsii*, has long been used by some Native Americans in religious ceremonies for its hallucinogenic properties.

Cacti are now widely cultivated as ornamentals. The climatic adaptability of cacti and their ease of culture make them useful in gardens and landscaping. Their unusual forms and spectacular, multicolored flowers, which vary in shade from green and white to magenta and purple, attract many collectors. It is never wise, however, to collect cacti from the wild. In many cases it is extremely difficult to duplicate an appropriate habitat and the cacti die. In other cases, collection of cactus species is illegal according to state or federal law.

Introduction   XV

Many nurseries carry a wide variety of inexpensive cacti and have nurtured the plants properly for the hobbyist. To seek the support from an established nursery is an environmentally wise and economically sound plan.

The following text presents cacti of Texas in a new perspective. We hope that you receive great enjoyment from this work.

Texas Cacti

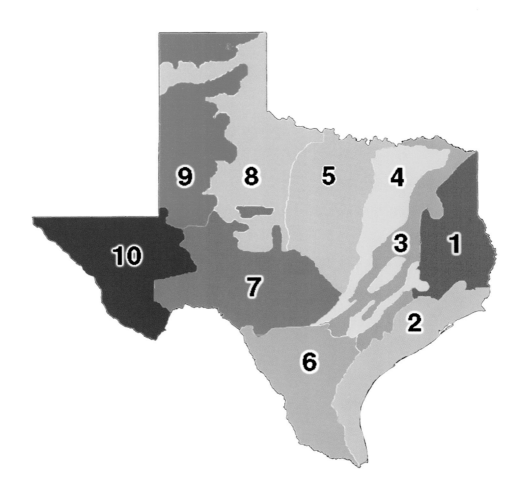

1. Pineywoods
2. Gulf Prairies and Marshes
3. Post Oak Savannah
4. Blackland Prairies
5. Cross Timbers and Prairies
6. South Texas Plains
7. Edwards Plateau
8. Rolling Plains
9. High Plains
10. Trans-Pecos

Cactus Country

Habitat

Cacti are known primarily as a New World family of plants, although in modern times many cactus communities have been established by human distribution in Africa and elsewhere on all continents except Antarctica.

In North America, cacti occur infrequently in Canada and abundantly through most of the temperate regions of the United States and well into Mexico. Cacti are found in greatest numbers and variation along the southern latitudes of the United States near the subtropical areas that extend southward from approximately 30° North latitude—between San Antonio and Austin, Texas—on farther south to the Tropic of Cancer near Ciudad Victoria in Tamaulipas, Mexico. Texas cacti are found in the greatest abundance in the Trans-Pecos region of West Texas and in the Rio Grande Valley.

Vegetational Areas of Texas

The state of Texas is vast. Understandably, the fauna and flora found within the borders are as diverse as the state is vast. However, within the state lie several areas, each with somewhat unique habitat, geographic diversity, and growth opportunities. These vegetational areas have been delineated and named as described below to better provide a depiction of ecological growth. This text will use these areas in reference to cactus distribution.

Several floristic associations also occur within cactus country. Floristic associations may vary within any vegetational area. Some cactus species are found only within a single floristic region. The most important of those are described within the area descriptions.

Differences in soil, climate, and rainfall are responsible for the great variety of vegetation found in Texas. Texas is home to all types of terrain, from forests in the east to semiarid, desertlike conditions in the west. Descriptions of the ten vegetational areas of Texas, from east to west, follow.

1. Pineywoods

The Pineywoods is an area of about 25,000 square miles that lies east of a line from the Texas-Arkansas state border at Texarkana, southwest to Longview and Crockett, then south to the Woodlands and east to Orange.

Elevation ranges from 200 to 500 feet above sea level. The area is nearly level to gently undulating and hilly. Many streams join into several large rivers and bayous. The area receives about 40 to 60 inches of rainfall fairly uniformly distributed throughout the year. Humidity and temperatures are typically high.

The soils of the region are of two types. Upland soils are generally acidic, sandy loams, sands, and clay. Bottomland soils are acidic to calcareous loam and alluvial clay.

The dominant vegetation type is a mixed pine-hardwood forest along streams and rivers with native pines in the uplands. Many species of shrubs, vines, forbs, and grasses occupy the forest floor, prairies, and areas not used for cropland. Several distinct floristic communities are within this forest region. The prominent is the Oak-Pine Woodland with its unique vegetative community.

Few cacti inhabit the Pineywoods. Those found in this area include *Escobaria missouriensis, Opuntia humifusa,* and *O. macrorhiza.*

2. Gulf Prairies and Marshes

The Gulf Prairies and Marshes area covers some 16,000 square miles, lies south of the Pineywoods, and spreads southeast of a line from Orange, west to Houston, southwest to near Beeville and Alice, and on south to Brownsville.

The Gulf Marshes are on a narrow strip adjacent to the coast and the barrier islands from Louisiana to Mexico. The Gulf Prairies are nearly level lowlands extending 30 to 80 miles inland from the marshes. Elevation ranges from 50 to 500 feet above sea level. They have slow surface drainage and are sliced by streams that flow into the Gulf of Mexico.

Rainfall in the area ranges from 26 to 56 inches annually.

Soils of the area are poorly drained sands, sandy loams, and clay. A narrow band of acidic sand, loam, and clay soils stretches along the coast. The loamy and clayey inland soils are commonly salty and basic. Prairie soils are neutral to slightly acidic loams and clays. Soils of the bottomlands and deltas are slightly acidic loams and clays.

The Gulf Marshes are a low, coastal area, commonly covered with brackish water, and rise to just a few feet in elevation. This area is divided into two regions, the marsh and salt grasses immediately at tidewater, and

farther inland, a strip of mixed short and tall grasses. Few hardwoods grow in the area. Vegetation types of the Gulf Prairies were historically tallgrass prairie and post oak savannah. However, communities of trees and shrubs such as honey mesquite, oaks, and acacia have increased and have become thickets in many places.

Several distinct floristic communities are within this region. They understandably include Coastal Plains and Coastal Barrier Islands and Dunes, each with its unique vegetative community. Within a very limited geographic area at the tip of Texas near Brownsville exists a remnant of Caribbean Tropical Forest. The presence of tropical species in this area is important to note. The few tropical species remaining are relics of this region that have been forced farther southward by climatic changes.

Cacti found in the coastal plains include *Escobaria missouriensis, Opuntia humifusa, O. leptocaulis,* and *O. macrorhiza. Opuntia pusilla* and *O. stricta* are found only on the barrier islands near Galveston.

Some cacti species are unique to the Caribbean Tropical Forest. These include *Acanthocereus tetragonus* and *Echinocereus berlandieri. Pereskia aculeata* and *Selenicereus spinulosus* are attributed to this area, but their current presence cannot be confirmed. Several species of prickly pear and chollas are also common throughout the area.

3. Post Oak Savannah

The Post Oak Savannah region is a narrow strip of about 14,000 square miles bounded by the Pineywoods and Gulf Prairies and Marshes on the east and a line roughly from Mt. Pleasant, south to San Antonio on the west. With its location in north-central Texas, this region is a transition zone between the Pineywoods and the western plains and associated prairies.

The terrain varies from nearly level to gently rolling hills and changes in elevation from 300 to 800 feet above sea level. The region receives an annual average rainfall of about 30 to 45 inches.

Upland soils are slightly acidic sandy loams, commonly over firm clays. The soils are frequently dry with clay pans at varying depths, which restricts moisture penetration. The bottomland soils are slightly acidic to calcareous loams and clays.

The region can be described as oak savannah, where patches of short oak woodland are interspersed with grassland. Hardwood trees are now common, along with native grasses and brush. The absence of recurring fires and other methods of woody plant control results in the development of thickets of post oak and blackjack oak.

Few species of cacti are found in this area of Texas. Species include *Cylindropuntia leptocaulis, Escobaria missouriensis, Opuntia humifusa,* and *O. macrorhiza.*

4. Blackland Prairies

The Blackland Prairies region is a narrow strip covering approximately 17,000 square miles bounded by the Post Oak Savannah on the east and a line roughly from Denison, south to San Antonio on the west. The terrain changes from nearly level to rolling hills with an elevation change from 250 to 700 feet above sea level. The area receives annual average rainfall of 30 to 45 inches.

The upland soils are dark clays. Bottomland soils are generally slightly acidic loams and clays. Soils here are generally fertile, but many have lost productivity through erosion and continuous cultivation.

The Blackland Prairies region is home to a variety of hardwoods and grasses. This historical tallgrass prairie was dominated by the big four prairie grasses as well as dropseeds and buffalograss. Today, much of the land is cultivated, and little original vegetation still remains. However, mesquite, huisache, oak, and elm are common species on poor rangeland and abandoned cropland. Elsewhere, oak, elm, cottonwood, and pecan are common along drainages.

The Blackland Prairies area intertwines with the Post Oak Savannah in the southeast. Together, this rolling prairie represents the southern reaches of true prairie that occurs

from Canada to Texas. Although most of the Blackland Prairies have been cultivated, small acreages of meadow remain as tallgrass prairie and forbs.

Cacti found here include *Coryphantha sulcata, Echinocactus texensis, Echinocereus coccineus, E. reichenbachii, Escobaria vivipara,* and several species of *Opuntia*.

5. Cross Timbers and Prairies

The Cross Timbers and Prairies area is nearly 26,000 square miles and lies west of the Blackland Prairies approximately from Denison on the northeast, west to Wichita Falls, then south to Brownwood, and southeast to San Antonio. This is an area of alternating woodlands and prairies with gently rolling hills that range in elevation from 500 to 1,500 feet above sea level. The area receives about 25 to 35 inches of rain per year.

Woodland upland soils are slightly acidic sands, sandy loams, and clay with limestone fragments. Bottomland soils have dark clays, and loamy alluvial soils occur along streams. The prairies are scattered with rapidly drained sandstone and shale ridges and hills. Prairie uplands are slightly acidic sandy loam over neutral to alkaline clay.

Some hardwood trees grow here. However, cultivation and poor past management have caused the uplands to be covered mostly by scrub oak, mesquite, and juniper with a variety of native prairie grasses. Bottomlands include hardwoods such as pecan, oak, and elm with recent invasion by mesquite.

Cacti found here include *Coryphantha sulcata, Echinocactus texensis, Echinocereus coccineus, E. reichenbachii, Escobaria missouriensis, E. vivipara,* and several species of *Opuntia.*

6. South Texas Plains

The South Texas Plains area covers about 31,000 square miles found south of a line from San Antonio to Del Rio and east to the Gulf Prairies and Marshes. This area is a transition from the Gulf Prairies and Marshes on the east and the plains of Mexico on the west. The area's nearly level to rolling terrain varies from sea level to 1,000 feet. Although the area suffers from increasingly frequent droughts, annual rainfall averages 16 to 35 inches.

Upland soils are of three types: dark and firm clays, loams, and sandy soils. Bottomlands are typically calcareous silty loams and alluvial clay soils.

The original vegetation was an open grassland or savannah along the near-coastal areas and brushy chaparral grassland in the uplands. Oaks and mesquite and other brushy species formed dense thickets on the ridges; and oak, pecan, and ash were common along streams. Continued grazing and cessation of fires altered the vegetation to such a degree that the region is now commonly called the South Texas Brush Country. This area has plains of thorny shrubs and trees and scattered patches of palms and subtropical woodlands in the Rio Grande Valley. The primary vegetation consists of thorny brush such as mesquite, acacia, and prickly pear mixed with areas of grassland. Because the South Texas Plains lie almost entirely below the freezing line, introduced tropical species do well in the subtropical environment.

The South Texas Plains region includes the species-rich Tamaulipan Biotic Province that extends southward approximately 200 miles from the Texas Hill Country into northern Mexico. The region encompasses some 17,000 square miles in Texas alone. The habitat known as Tamaulipan thorn scrub is dominated by a diversity of woody plants, including honey mesquite, various acacias, post oak and live oak, graneneo, cenizo, and whitebrush. Because of the rich mixture of diverse biological elements, many groups of flora and fauna are represented in greater numbers than in any other similar North American regions. The Lower Rio Grande Valley is known as the most biologically diverse region in the United States.

Because of this biodiversity, this is one of the most important areas rich in cacti within the United States, and there are many endemic species.

Cacti include many prickly pears and chollas; however, many *Coryphantha, Echinocactus, Echinocereus, Mammillaria,* and *Sclerocactus* species do well here, including *Echinocereus berlandieri, E. pentalophus,* and *E. poselgeri.* Several threatened, endangered, and protected cacti are also found in this area, including *Astrophytum asterias, Coryphantha macromeris* var. *runyonii,* and *Lophophora williamsii.*

7. Edwards Plateau

The Edwards Plateau area includes about 37,000 square miles found west of a line from San Antonio to Midland, then south to Fort Stockton and the Rio Grande roughly at the Brewster County line, then southeast to Del Rio, and back to San Antonio.

The terrain is varied from rolling, hilly, to mountainous terrain, including woodlands in the west and stony prairies in the east. Elevation ranges from about 800 to 3,000 feet above sea level. Average annual rainfall ranges from 15 to 34 inches.

The area is a rapidly drained stony plain deeply cut by streams and rivers, creating flat to broad, undulating divides. Soil composition is shallow and limestone based. Upland soils are calcareous clays and loams that are mostly stony. Bottomland soils include areas of calcareous clays and alluvial soils.

The original vegetation was grassland or open savannahs with tree or brushy species found along rocky slopes and stream bottoms. Tall grasses are still common along rocky outcrops and protected areas having good soil moisture. These species have been replaced on shallow xeric sites by short grasses. The western part of the area comprises the Stockton Plateau, which is more arid and supports short- to midgrass mixed vegetation.

The Edwards Plateau region includes an area commonly known as the Texas Hill Country. It is a region of many springs, stony hills, and steep canyons. The Edwards Plateau is characterized as oak-juniper woodland where mature ashe junipers (cedar) and various oaks make up the majority of this area. Predominant woody species include live oak, shin oak, honey mesquite, Mexican persimmon, hackberry, Texas ash, and bald cypress.

The region is home to an abundance of rare plants and animals found nowhere else. Cacti are abundant, including species of *Coryphantha, Echinocereus, Mammillaria*, and one endemic, *Sclerocactus brevihamatus* var. *tobuschii*. Chollas and prickly pears are common on overgrazed ranges.

8. Rolling Plains

The Rolling Plains area includes about 37,000 square miles, lies immediately north of the Edwards Plateau, and covers most of the Texas Panhandle along with the High Plains. It is roughly bounded by a line from Wichita Falls, west to Amarillo, south to Lubbock and Midland, east to Abilene, and back north to Wichita Falls.

It is a nearly level to rolling plain with moderate to rapid surface drainage. Elevation varies from 1,000 to 3,000 feet above sea level. Annual rainfall averages 18 to 28 inches.

Soils range from coarse sands to tight or compact clays. Soils of the uplands are calcareous sandy loams and clays. Saline soils are common, as are shallow and stony soils with pockets of deep sand. Bottomlands have loam and clay-based calcareous alluvial soils.

The area is half mesquite woodland and half prairie. The original prairie vegetation included mixed tall and midgrasses. Short grasses are more common on the more xeric or overgrazed sites. Continued overgrazing and reduction of fires have created change in habitat to short grass, shrubs, and annuals. Mesquite, lotebush, prickly pear, agerita, and tasajillo are common on all soils. Shin oak and sand sagebrush are found on the sandy lands, and redberry juniper has spread from rocky slopes to grassland areas.

Several unclassified grasslands occur in Texas. They are modified from original prairie grassland and are transitional between Great Plains Grasslands and Desert Grasslands.

This area is not known to be rich in cacti. As with other areas in Texas, prickly pears and chollas are found with frequency. *Coryphantha* and *Echinocereus* species are infrequent.

9. High Plains

The High Plains, an area of about 31,000 square miles, is part of the Great Plains and occupies the balance of the Panhandle, separated from the Rolling Plains by the Llano Estacado Escarpment and divided by the Canadian River.

The area is a nearly level high plateau where the elevation changes from 3,000 to 4,500 feet above sea level. Rainfall in this part of Texas is relatively modest, from 14 to 21 inches annually on average. However, this plateau contains many shallow, silty depressions, or playa lakes, which sometimes cover as much as 40 acres and contain several feet of water after infrequent heavy rains. These depressions support unique patterns of vegetation within their confines.

Soils are mostly clay, loam, sandy loam, and sands. High wind, dry winters, and low rainfall present problems for cultivation and create detrimental erosion. The canyons of the Caprock Escarpment with their impressive walls were formed by this action and form the headwaters of the Red River.

The High Plains region is a mixed prairie with virtually no trees. The original vegetation was classified as mixed shortgrass and tallgrass prairie. Gramas and buffalograss are the principal vegetation on the clay and clay loam sites. Shin oak and sand sagebrush are found on sandy sites. The High Plains area is historically free of brush, but sand sagebrush and western honey mesquite, along with prickly pear and yucca, commonly invade the sandy and loamy sites.

This, too, is an area not known to be rich in cacti. However, prickly pears and chollas are frequently found. *Coryphantha* and *Echinocereus* species are sometimes encountered.

10. Trans-Pecos

The Trans-Pecos, a region covering 28,000 square miles, makes up the corner of Texas generally west of the Pecos River and Fort Stockton and on to El Paso. This is the West Texas of legend, commonly pictured as inhospitable rocky lands of desert, mountains, mesas, canyons, flat basins, and arid valleys.

The Trans-Pecos is the most complex of all the regions. Diverse habitats and vegetation vary from desert valleys and plateaus to wooded mountain slopes. Elevation ranges from 2,500 feet to more than 8,749 feet above sea level at Guadalupe Peak. Mountains are varied, some characterized by volcanic rocks, others by limestone.

With as little as 8 to 18 inches of annual rain and long, hot summers,

Southwestern chaparral vegetation is common in the Trans-Pecos and the Chihuahuan Desert.

only vegetation adapted to desertlike conditions can flourish in the basins. Over most of the area average annual rainfall varies greatly from year to year and fluctuates considerably with change in elevation.

Impressive canyons cut the land into significant drainages where mountain outwash materials have formed the varied soil textures and characteristics of the Trans-Pecos. Typically, alkaline soils with high salt content and gypsum dunes are common in the basin areas; upland soils are mostly clays, loams, and sands over calcareous or gypsum subsoil.

The original vegetation ranged from desert grassland and desert shrub on lower slopes to juniper and pinyon pine at mid-elevations. The mountains support ponderosa pine and forest vegetation at higher elevations. Vegetation, especially on the higher mountain slopes, includes many southwestern and Rocky Mountain species not present elsewhere in Texas.

Plains and desert grasslands are found in the Trans-Pecos. Plains grasslands are found on the lower mountain elevations and in the outwash basins. Desert grasslands occur on plateaus, rolling hills, and basin floors with relatively deep soils. Grass flats occur in low areas where water runoff accumulates, while a variety of yuccas commonly inhabit hillsides with improved drainage.

Lower elevations of the region are characterized as Chihuahuan desert scrub, comprising up to one-half of the vegetation in the region. Creosotebush is a prominent species along with leaf succulents such as lechuguilla, sotol, and other yuccas. These are dominant indicator plants of the desert scrub landscape. Other common shrubs include acacia, mesquite, saltbush, javelinabush, allthorn, and ocotillo. Without the previous cover of perennial grass, the area is subject to erosion from intense summer thunderstorms.

Southwestern chaparral consists of brushland dispersed through an area of oak woodlands, characterized by only a few shrubby species. A typical chaparral plant community consists of densely growing evergreen scrub oaks and other drought-resistant shrubs. Chaparral occurs on steep hillsides with poor, thin soil that cannot support larger plants.

Juniper-pinyon woodlands characterize the slopes and valleys at mid-elevations. Common woodland trees include oaks, junipers, pines, madrone, and bigtooth maple. Woodland shrubs include mountain laurel and mountain sage.

This is one of the most cactus-rich areas of the United States. Cacti are mainly small but make up for size in their variety and numbers. While the Chihuahuan Desert flora is best represented in Mexico, the best populations of these cacti in the United States occur in the Big Bend country of the Rio Grande. The Trans-Pecos is home to more than 120 recorded cactus species, subspecies, and varieties, including endemics such as *Cylindropuntia imbricata* var. *argentea, Echinocereus chisosensis, E. viridiflorus* var. *davisii, Escobaria hesteri, E. minima, Opuntia aureispina*, and *O. chisosensis.*

Cactus Anatomy

What Is a Cactus?

A cactus is a succulent, perennial, vascular plant. The name "cactus" has been misused as the universal term to describe succulent plants, those fleshy species that have the capacity to store water within their tissues and to prevent its loss during dry periods. Often the term has been misused to describe a varied range of plants, including aloes, agaves, euphorbia, ocotillos, and yuccas.

Many succulents look like cacti but are not. It is often said that all cacti are succulents but all succulents are not cacti. The true cacti are members of the botanical family Cactaceae. They are distinguished from other succulents and from other vascular plant genera by several anatomical structures, including the following:

Areoles: A meristematic, or "growth tissue," structure found on cactus stems, giving rise to flowers, spines, other stems, and sometimes roots. Areoles are basically axillary buds similar to those of other plants, but they are highly modified in cacti.

Flowers: Flowers are not unique to cacti, but cacti flowers are unique. Cactus flowers are typically quite spectacular and very complex with a unique anatomical structure.

Leaves: Most cacti do not have true leaves; however, some do. Frequently cacti have ephemeral leaves found only on young growth.

Origin: Most all cacti are native to the Americas and surrounding islands, with few exceptions.

Worldwide cultivation now makes this less helpful for identification.

Spines: The most familiar feature of cacti. Spines can vary greatly in appearance, shape, size, physical arrangement, and color.

Several evolutionary changes in cacti include three most important modifications: the loss of leaves, which reduces the surface area of the plant and associated water loss; the growth of the stem into water-storage tissue; and the adaptation of leaves into spines, which help to protect stored water. These and other changes are important in cactus anatomy and will be further discussed in the sections appropriate to these features.

Features of Cacti

Texas cacti vary greatly in size and general appearance, from tiny, inconspicuous buttonlike *Epithelantha,* to low, clumping hedgehog *Echinocereus,* to spreading shrubs of *Opuntia* prickly pears, and the massive upright columns of the barrel *Ferocactus.* Many cacti grow low in the ground, others grow into large treelike structures, and some clamber and thrive only with the protection and support of a host plant.

A cactus plant is conspicuous for its fleshy, green, chlorophyll-containing stems that perform the functions of leaves. In most species leaves are either absent or greatly reduced, minimizing the amount of surface area from which water can be lost. Most often conspicuous are the spines of various colors, shapes, numbers, and arrangements. Cactus flowers can be large and showy; are commonly yellow, white, or shades of red and purple; and possess an anatomical structure that is unique in the world of vascular plants.

These characteristics of cacti make them unusual among the plant world and provide a need for some more in-depth understanding. Let us take a look.

Roots

Most Texas cacti produce many fine, multiple-branching roots that spread quickly just beneath the soil surface. The roots of these plants are shallow; they may penetrate the soil only a few inches but may cover a large horizontal area. These roots quickly absorb any available moisture that has penetrated the soil following light, infrequent rains. Shallow-rooted cacti also do not need to compete with other, more deeply rooted species in the same habitat.

Some species have more compact systems, designed to acquire their moisture, perhaps even from the plants themselves, as water drips from the spines to be absorbed by the shallow roots. The plants then store this water to the limit of their capacity. Some cacti may have a more compact root system of short lateral roots just under the soil surface. These

Cactus Anatomy

Cactus spines provide an ideal surface for atmospheric condensation and allow that water as well as rainwater to collect and drop near the base of the plant to be absorbed by its shallow root system.

Lace cactus is an example of a rather simple cactus with columnar stems and infrequent branching near the stem apex.

roots allow rapid absorption of water that runs off from the plants themselves after a rain or from overnight condensation on the spines.

Most cacti store water in the shoot, but a number of cacti store water in large, succulent taproots with small secondary roots arising from them. In these plants, water is stored underground where it is out of sight of thirsty animals, it is cooler than in the air, and its weight does not have to be supported.

Some slender, climbing species have massive underground roots that appear like turnip-shaped taproots. Climbing and creeping cacti may produce roots at many points where the stems contact the soil.

Stems

The most obvious feature of cacti is the stem, or main body of the plant. In most cactus genera mature stems have only microscopically tiny true leaves, revealing a large, spiny aboveground structure representing the majority of the plant mass.

The cactus stem and its features may vary greatly, adding to the wide diversity of forms of cacti. The stems of cacti may be simple, unbranched, and rounded or columnar. They may also be branched either at the ground level or many times well above it. In some genera, a series of stem segments with joints may be present, as in the chollas and prickly pears.

These structural features are an important component to cactus identification.

Epidermis

The majority of the biomass of a cactus is its stem. Having no leaves, the cactus stem must take over the leaf functions, including photosynthesis. The stem is covered by a thick, waxy skin, or epidermis, which is impervious to water. It is equipped with specialized cells that allow the selected passage of water and gas into the tissue. These pores, or stomata, also close when needed to reduce the loss of water.

Before leaves could be eliminated from its anatomy over time, the plant stem itself had to become photosynthetic. That required the cactus epidermis to become long-lived with a high density of stomata.

Because the epidermis of the stem of a cactus is crucial to desert survival, it must not be shed, so an unusual feature of cacti is that they produce their cork cambium and bark only after they have become extremely old. The cork cambium arises from the epidermis. This surface location ensures that the cortex is retained forever.

Cortex

It is easy to take for granted the thick cortex of cacti, the layer of tissue just beneath the epidermis. It is unique; no other group of plants has such a thick cortex. The cactus cortex is thick because this is the location of a system of cortical bundles that transfers water, minerals, and organic nutrients throughout its extensive volume. The cortex also has modifications that allow it to be photosynthetic, to swell and contract without tearing apart, and to have water-storage cells.

The vascular tissue of cactus consists primarily of two types. One transports water from the roots;

The green color in the pads of this prickly pear indicates that living epidermis is present on all parts of its stem, even parts that are many years old.

Cactus Anatomy

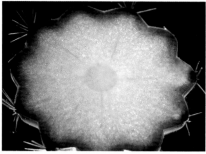 

Thin cross-section slices of a typical *Echinocereus* stem. (Left) Freshly cut slice shows areoles and spines, outer epidermis, a dark green photosynthetic cortex layer, and central pith. (Right) Stained section shows a dark ring of vascular bundles near the center, radiating leaf/bud traces, and numerous other threads that are cortical vascular bundles.

another transports food, or sugars, throughout the plant. As plants age, the vascular tissue produces wood.

Cactus wood transports water, and in some plants it also supports the plant's weight. But globose cacti such as barrel cacti are supported by their internal water pressure, not by wood. In many cacti, wood fibers have been reduced or eliminated, and instead the volume of water-storage tissue is very high. Most cacti also have elastic cellular structures allowing the tissues to shrink during drought, preventing the formation of empty cavities—a main cause of water stress—as the plant's volume shrinks to accommodate the water volume within it.

Stem Joints

Jointed stem segments are commonly characteristic of opuntias, the prickly pears and chollas. A joint is a shoot segment, a stem that arises from an areole. The jointed growth of the plant may follow a relatively straight line or may be very random in direction. This segmentation by jointing is a method of prolific vegetative reproduction, as the segments are broken off and mechanically dispersed.

Pads of opuntias are actually jointed stems.

Tubercles

The surface of cacti may be smooth, but often they are covered with enlarged nipple-shaped structures called tubercles. Tubercles are actually enlarged leaf bases bearing an areole, or specialized growth area giving rise to spines or other anatomical structures.

The tubercles on this *Escobaria* are readily identified as raised conical structures of stem tissue with areoles and spines on each apex.

Distinct ribs of this *Ferocactus* clearly demonstrate this plant's ability to expand as it absorbs water.

Ribs

In some species the tubercles become confluent, producing longitudinal ridges or ribs running vertically or spirally along the cactus stem.

These tubercles and the ribs in columnar cacti provide the stem an accordion-like expansion capability that allows the cactus to store water during times of rainfall. This capability allows swelling without tearing. In times of drought, these structures will contract with the loss of tissue fluids.

Because ribs provide the cactus stem a somewhat folded surface, they also provide additional strength, flexibility, and longitudinal stability. Tubercles and ribs are important identifying features of Texas cacti and should be noted with detail in the field.

Apical Growth

As cacti became more succulent and heavier through evolution, the number of branches per plant was reduced. Many cacti now have one stem with only one shoot apical growth area (meristem). Whereas an ordinary tree is constructed by the activity of thousands of shoot apical meristems, cacti are produced by just a few or only one. The result is that the number of cell divisions that each cactus apical meristem cell must undergo could be very high—this could lead to increased risk of mutations.

Areoles

Cactus areoles are unique, important structures that separate the cacti from other succulents. Areoles are a derivative of the bud formed in other vascular plants above the position of a leaf on a stem. Areoles are highly specialized vegetative tissues or short shoots that may give rise to the formation of leaves, spines, hairs, flowers, fruits, or other stems. In certain circumstances the areole may also give rise to the formation of roots.

Areoles have three basic shapes. Circular or oval areoles are the most primitive and most common. Elongated and two-part areoles are also found, especially in tuberculate cacti. Often the two-part areole is connected by a groove.

Areoles are essentially bilaterally symmetrical. Spines generally arise from the edges. In some cacti, however, the elongated areoles have functions differentiated by position; the distal portion produces flowers and stems, and the more basal portion produces spines.

Spines

Among the most distinctive features of cacti are the spines that arise from areoles. Only in cacti do stems bear spines in clusters. Cactus spines are nonvascularized, modified leaves that develop from the meristematic tissue within the areole.

Like leaves of other vascular plants, spines vary in number, physical

These photos clearly illustrate differences among cactus areoles. (Top to Bottom) Peyote with spineless, hair-filled areoles; Arizona barrel with its large, two-part areoles; and lace cactus with its elongated oval areoles.

Arising from an areole in clusters, cactus spines are unique and distinctive features. Their shape and growth patterns are useful in plant identification.

arrangement, size, shape, and cross section. As they grow, spines harden from the tip downward as a cuticle is deposited on its outer surface. Spines may also vary in color due to the deposition of water-soluble pigments. These colored spines may also be opaque or translucent.

Spines seem to provide several functions; the most obvious and painful is a mechanical protective defense against animal consumption. A secondary function is to cause portions of the stem itself to be transported in the hairs of animals to aid in its reproductive dispersal.

In some species the spines are projected or curved downward, directing the drip of condensation or rainwater droplets toward the base of the stem

Extrafloral nectaries may be identified as small, reddish swellings (arrows) in the areoles of *Sclerocactus brevihamatus*.

Simple spinal arrangement of *Echinocereus coccineus* features a single central spine and eight radial spines arising from an oval areole on the distinct rib.

and its roots. Spines of some species are also many and feathery, obscuring the stem and shading the plant from severe sunlight. Spines of some cacti even mimic blades of dry grass, effectively camouflaging the plant.

In some species spines have become secretory glands, or extrafloral nectaries, producing sweet, sugar-based nectar. The nectaries are actually modified spines with loose fiber bundles that conduct plant fluids to the areole surface by capillary action. The nectaries attract pollinators and ants alike. There is some disagreement among researchers about whether the ants and insects visiting the extrafloral nectaries are part of a symbiotic relationship that aids in cactus pollination.

Spines are frequently a significant aid to species identification. Within the areole there are spines situated more centrally and spines located more around the edges of the structure. It is the number, arrangement, and features of these central and radial spines that lead us to confirm the identification of a cactus species.

Glochids

Glochids are unusually modified spines found only in the opuntias. They are shorter and much thinner than the larger spines and, in addition to traditional spines, occur in great numbers in the areoles of these species. Unlike spines, glochids are brittle. They separate from their base and become easily dislodged from the

Glochids fill the areoles on the edges of the pads of many opuntias, as these highly magnified views of *Opuntia engelmannii* demonstrate (left, 10×). The individual glochid (right, 40×) shows the barbed surface composed of many stacked cellular plates.

areole. Very thin, sharp, and barbed, glochids are obnoxiously painful and very difficult to remove when lodged in human skin.

Leaves

In outward appearance cacti do not have leaves. While most cacti lack typical flat, green leaves for photosynthesis, all cacti still produce leaves. Some species have remarkably large, ordinary leaves, but most have only tiny, often ephemeral ones. The loss of leaves not only reduces the surface area through which cacti lose water but also eliminates the extensive network of leaf veins through which water is transported.

Leaves persist in a few species of cacti throughout the life of the plant. True leaves and spines are borne on the stems of *Pereskia*. Although pereskias appear unlike most common cacti in form, some researchers believe they are very similar to the earliest types of cacti from which all others evolved.

Some people mistake the large, rounded pads of prickly pear cacti as leaves; however, the pads are actually flattened stem segments that bear areoles with spines, flowers, and fruits. Therefore, the majority of cactus species do not produce typical green leaves but have produced in their place the highly specialized spines characteristic of most cacti.

Ephemeral leaves of an *Opuntia* are clearly visible on the left. They will wither and fall off while only a few weeks old. True leaves of a *Pereskia,* a cactus with persistent photosynthetic leaves, are illustrated on the right.

Flowers

Cacti produce remarkably stunning flowers. They are generally large, colorful, and very showy. Cactus flowers, as well as their interesting shapes and textures, are a great inspiration that drives cactus enthusiasts and stimulates the destructive collecting of cacti. While colorful and showy, these flowers are equally complex. A flower is borne on an areole, which is actually a specialized short shoot.

Cactus flowers are one of the distinguishing features of cacti. The flowers have a number of structural characteristics important to taxonomists as an aid to cactus identification.

Cactus flowers are "inside out." They are hollow tubes, as shown in the image on p.26, with petals at the top, stamens inserted on the inside of the tube, and carpels at the very bottom. The whole flower is embedded in shoot tissue that bears true leaves called bracts.

Cactus flowers are distinguished by the presence of an inferior ovary—that is, the ovary is located beneath the other parts of the cactus flower. Cactus flowers come in all colors except blue.

Pollen

Like other anatomical structures, cactus pollen grains vary considerably among species and provide valuable information on plant relationships. Cacti are actively pollinated by a variety of animal agents such as bats, birds, and insects, including bees, butterflies, and perhaps ants.

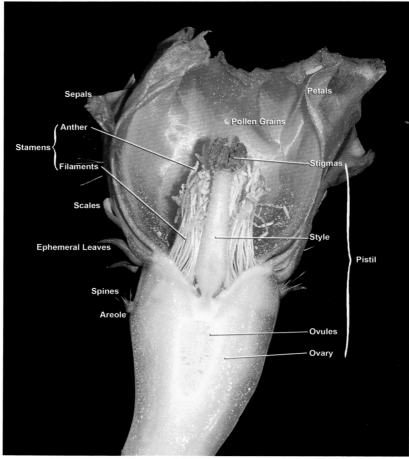

Longitudinal section of a prickly pear cactus is representative of the floral anatomy of many cactus species.

Fruits

The cactus fruit, called a tuna in some species of opuntias, matures from the floral ovary. It is somewhat protected by spines, glochids, or wool. It grows to become a berry with many small seeds. The berry dries and splits, and the seeds fall out. In some cases the fruit becomes fleshy, loses its spines, and drops off the stem. Birds and animals carry the fruit off and effect the seed distribution; thus, the plant becomes widely dispersed.

Some cactus fruits are sweet; candies, jellies, and preserves are often made from the fleshy components. However, beware the jelly maker who does not take appropriate care to remove all the spines and glochids!

Cactus Anatomy

Images of cactus bud, flower, and spent flower. The images above demonstrate the morphological changes in a cactus flower throughout its season. Left is a developing bud. Note the residual ephemeral leaves and glochids. Also visible are the specialized petals, or bracts. The flower is fully formed in the center. The right image is of the spent, collapsing flower. Soon the flower will fall, leaving only its scar (umbilicus) on the inferior ovary, which will ripen into the fruit, as demonstrated in the image, below.

Cactus fruits have many forms. Purple, ripened fruits of a prickly pear cactus, called tunas (left), may be used for a variety of foods. The light gray circles are the umbilical scars remaining where the flowers were once attached. These fruits measure about 5 cm (2 in) across and contain about 400 seeds. Red, ripened fruits of *Mammillaria lasiacantha* are only 1 cm (1/2 in) long and take on a completely different appearance. In contrast, they may contain only 15–20 seeds.

Fruits of *Opuntia* and other genera may contain hundreds of seeds. The fruits will ripen, dry, and split. The exposed seeds will then disperse. In some cases mammals, birds, and even ants tend to accelerate this process. In the photo above, the tuna has been pecked open by a bird.

Seeds

Seeds are important in understanding the relationship of cacti. There is much variation in size, shape, color, and microscopic surface pattern of the seed. There are two basic types of seed shape among cacti within Texas. The opuntia type includes seeds from chollas and prickly pears and is basically flattened and disklike, yellowish or cream colored, very hard, and about the size of a small pea. The cereus type of other genera includes many modifications in form, from ovoid or obovoid, to nearly spherical, to many other irregular shapes and surface characteristics; they are frequently black or brown and tiny, about the size of a poppy seed.

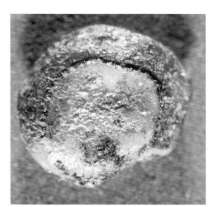

Opuntia seeds (left, 16×) are a flattened disk. The cereus type (right, 50×) is frequently black and helmet-shaped. Surface textures vary from smooth to papillate.

Cactus Critters

Not surprisingly, animal life abounds in cactus country. Even the desert is not devoid of its many varieties of animals. Some are casual inhabitants; many others, more specific to this habitat, have an interesting relationship with cacti. A visitor exploring cactus habitat may find many of the creatures that dwell among the cacti.

More than two hundred species of ants occur in cactus country, ranging in size from less than one-sixteenth to more than an inch long. Ants are omnivorous and frequent cacti in search of food, where they may accidentally pollinate the blossoms as they forage through the filaments.

Ants are also industrious and harvest seeds for later use, including those of cacti. Ants have been known to spread cacti over large areas through this method. Other, larger ants are hunters, stalking through the stems and spines of cacti in search of a variety of prey. Some cacti possess extrafloral nectaries that produce a sugar-water liquid often laced with amino acids. Ants are documented to visit these sugar sources to feed on the liquid nutrient.

The bee assassin is a variably colored insect that preys upon honey bees and other insect pollinators.

More common in the Southwest, these insects lie in wait, frequently within the large flowers of cacti and other plants. These insects pounce upon their prey, securing it with its strong legs. The assassin uses its hollow, hypodermic-like beak to pierce the victim, inject an immobilizing digestive enzyme, and then suck out the body fluids.

An insect common to prickly pear cactus is the leaffooted bug of the genus *Narnia*. This insect looks similar to an elongated stink bug with winglike expansions of the upper hind legs that give it its common name. These insects are equipped with a piercing proboscis and feed upon the juices of the plant.

Cochineal bugs are the most cryptic of insects. The females of these insects spend their entire life cycle in place concealed by a funguslike mass of dense strands of white cottony wax. Often colonies of these insects are conspicuous, covering pads of prickly pears like a white fluffy mat. The insects are actually red with deep red, waxy scales under their bodies. When crushed, these insects make a deep crimson stain, and this material has been used for the production of bright red, orange, and purple dyes. The production of this dye, used from Aztec times, once stimulated a large industry that flourished through cultivating prickly pear cacti and harvesting the insects for the dye. Synthetic dyes have now virtually eliminated this industry.

In cactus country even the most obscure and cryptic of cacti produce often spectacular flowers with the specific purpose of propagation through pollination. In cactus habitat, cryptic cactus plants may be more easily discovered by watching the bees. During flowering periods, honey

Cactus Critters

bees and other species abound and head to the cactus flowers with great regularity. Their antics are most interesting as they search for their prized pollen, seeming to wallow in the many filaments of blossoms. In larger flowers they may actually disappear from sight among the structures.

Nearly five hundred species of butterflies have been found in Texas. Combined with the larger population of moths and skippers, that number approaches sixteen hundred species. Butterflies and moths are occasional visitors to cactus flowers to drink nectar from the blossoms. Therefore, they become pollinators of unknown significance. In areas approaching Texas the cactus moth introduced from

Argentina is destructive to prickly pear cacti. Today it threatens the $100 million cactus industry in Mexico.

There are accounts of moth and cactus relationships recorded from Canada southward through Arizona and into Mexico. In Arizona the *Senita* cactus is the host plant to the senita moth, which is the sole pollinator of that cactus and thus is rewarded by its nectar and immature fruit, the only food of the moth. Eggs are laid within the *Senita* flower, where the flower is pollinated. The larvae move downward through the pollen tubes into the ovary to consume the unripe ovules. As they mature, they eat their way into the cactus stem, where they pupate.

The many species of jumping spiders are seemingly curious little arachnids and frequently have brightly colored pedipalps just below the mouthparts. These spiders hunt insects by stalking through plant vegetation and pouncing upon their prey. Although usually very small, some jumping spiders may inflict a painful, though nonpoisonous bite to humans.

A dark predator that raises fear and the hair of most people is the tarantula, the largest spider in Texas. Tarantulas are relatively common, and their large size and hairy appearance make them easily recognizable. Tarantulas may be found throughout the state in grasslands and other semiarid, open areas. Fourteen species occur in the state, and all may in-

flict a painful, yet nonpoisonous bite. These spiders predate upon other spiders, crickets, grasshoppers, caterpillars, and other insects. Ambushing their prey from their burrows or while foraging, they will inject a digestive enzyme into the body of the prey with their sizable fangs. Periodically a tarantula migration will occur during the summer, which may become an impressive sight.

Easily recognized by the set of pincers on the first pair of appendages and the upward curving tail with a stinger, the scorpion is a somewhat common inhabitant in the more arid regions of cactus country. Eighteen species of scorpions are found in Texas, many of which are uncommon. Scorpions are most readily found under loose, decaying bark and under rocks and ground rubble. They are most active at night. Ranging in size up to 3 inches, scorpions can inflict a painful sting. Contrary to popular belief however, Texas species are not considered deadly.

North America's largest wren, the cactus wren, occupies much of the southwestern United States from Southern California to South Texas and on into Mexico. Its habitat, as the name suggests, is limited to desert and semidesert areas abundant with larger cacti, including chollas and prickly pears. Cactus wrens derive their name from the habit of building their nest in the middle of larger cactus clumps amid the protection of spines. This football-shaped nest, with a side entrance and constructed mostly of grasses, is easily identified, as is the bird's unusual call, a distinctive sound of the desert.

Of all the feared desert creatures, the rattlesnake has the most wide-

spread reputation. Texas has nine species of rattlesnakes. The Western diamondback is the largest, most common, and widespread venomous snake in the state. This predominantly nocturnal snake spends most of the warmer daylight hours in the shade of larger plants, rocks, or earthen burrows made by other animals. Generally an ambush predator, the rattlesnake can detect its meal of small rodents and other mammals through the use of its sensory pits, detecting

the infrared emissions of the victim in a manner far more sensitive than by sight alone.

Cactus spines are of value to the plant for several reasons, one as a form of defense against physical damage by animals. However, some larger mammals can bite through the armor of cacti like prickly pear and actually consume the succulent tissue of the plant. While this practice may be uncommon, perhaps only in times of drought, javelina, deer, and cattle have been known to eat the pads in this manner. Because of this practice, botanists, led by Luther Burbank, once attempted to develop a form of spineless prickly pear. Although this fortune-making scheme was short-lived, a few stands of these cultivated plants can still be seen in Southern California.

How to Use This Book

The cactus species presented in this book have been conveniently arranged so that an enthusiast in the field can identify the plant in question without the use of a botanical key.

Cactus plants have an overall shape that may be used to separate one genus from another. In addition to the shape of the stem, the shape, growth pattern, and other physical characteristics of the spines are significant to the identification of the plant.

Therefore, the species within the species accounts have been separated by their most common stem shape. Most species accounts provide the reader with three detailed photographs: a large close-up to provide a clear view of the form and structure of the plant as a whole; an extreme close-up of the spines to demonstrate their shape and arrangement; and a habitat image that provides an appreciation of the plant's growth within its environment.

Each section within the accounts is separated by color-coded page edges and an accompanying stem-shape icon for ease of recognition. In addition, each species has two other icons associated with it to remind the reader of the blooming time and conservation status in Texas.

Range maps are provided with each species account to demonstrate the geographic area where that species may occur. These range maps are stylized graphically to best represent the predicted range. The reader must be aware that there may be areas within the predicted range where the cactus is not found, and conversely, individuals of that species may be found outside the indicated area. These maps were individually hand-crafted using *Cacti of the Trans-Pecos; Flora of North America,* Vol. 4; the *BONAP-TAMU Checklist of Vascular Flora;* and *The Atlas of Vascular Plants of Texas* as sources. The maps also incorporate geographic ranges from written descriptions in *Cacti of the Southwest* and a database provided by Texas Parks and Wildlife.

Where possible, complex terminology is kept to a minimum. However, it is difficult to describe some cactus features without the use of terms perhaps unfamiliar to many readers. The illustrated terminology chart will help the cactus enthusiast become familiar with some of these terms.

Cactus Stem Shapes

MAT-FORMING SEGMENTED
Corynopuntia

ROUND SEGMENTED
Cylindropuntia

FLAT SEGMENTED
Opuntia

ANGULAR
Acanthocereus
Echinocereus
Peniocereus

CLIMBING/CLAMBERING
Echinocereus

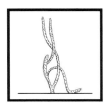

CYLINDRICAL
Coryphantha
Echinocereus
Escobaria
Mammillaria
Neolloydia
Sclerocactus
Thelocactus

LOW, SPINELESS
Ariocarpus
Astrophytum
Lophophora

STRONGLY RIBBED
Echinocactus
Ferocactus

GLOBULAR
Coryphantha
Echinocereus
Epithelantha
Escobaria
Mammillaria
Sclerocactus

How to Use This Book

Blooming Time
Day

Cacti that open their flowers during the daylight hours. The flowering period may be for only one day and the flowers do not reopen, or for several days in succession. Diurnal bloomers take advantage of daytime pollinators such as ants and bees for reproduction.

Night

Cacti that open their flowers late in the day or during the hours of darkness. Nocturnal bloomers take advantage of night pollinators such as moths and bats for reproduction.

Conservation Status in Texas

In Texas as well as elsewhere, cacti enjoy a special form of legal protection in their habitat from digging, taking, possession, transportation, or sale. The Endangered Species Act is federal legislation designed to provide a means to conserve the ecosystems on which endangered and threatened species depend and provides programs for the conservation of those species, thereby preventing extinction of plants and animals.

The law is administered by the United States Interior Department, Fish and Wildlife Service and within Texas by the Texas Parks and Wildlife Division. A species, subspecies, or distinct population segment is added to the federal list of endangered and threatened wildlife and plants through a formal process.

Within this process are several levels of concern for plant populations. They include endangered plant species, which are in danger of extinction throughout all or a significant portion of their range. Threatened species are likely to become endangered within the foreseeable future throughout all or a significant portion of their range. Of conservation concern are species that may require periodic monitoring of populations and threats to them and their habitat to assess the necessity for listing as threatened or endangered.

Within Texas, conservationists use the following system for ranking plant status. The cacti within this book are identified using this system as well. An icon will appear on each page within the species descriptions indicating each plant's status at the time of printing.

CRITICALLY IMPERILED
Critically imperiled in the state because of extreme rarity (often 5 or fewer occurrences) or because of some factor such as very steep declines, making it especially vulnerable to extirpation from the state.

IMPERILED
Imperiled in the state because of rarity due to very restricted range, very few population numbers (often 20 or fewer), steep declines, or other factors making it very vulnerable to extirpation from the state.

VULNERABLE
Vulnerable in the state due to a restricted range, relatively few population numbers (often 80 or fewer), recent and widespread declines, or other factors making it vulnerable to extirpation.

APPARENTLY SECURE
Uncommon but not rare; some cause for long-term concern due to declines or other factors.

SECURE
Common, widespread, and abundant in the state.

Cactus Terminology
Stems and Fruits

Clavate	Club shaped
Cylindroid	Cylinder shaped
Elliptic	A pointed oval about twice as long as wide
Globose	Rounded or spherical; globelike
Obovate	Egg shaped, with the widest part above the middle
Ovate	Egg shaped, with the widest part near the base

Spines
Diameter

Hairlike	Very fine in appearance
Slender	Moderate in cross section
Stout	Massive in appearance

Cross Section

Acicular	Needle shaped; long axis tapering to a point
Angled	Not rounded; usually triangular throughout the long axis
Flattened	Compressed in one dimension across the long axis
Terete	Cylindrical throughout the long axis

Surface Texture

Plumose	Featherlike
Ridged	Rings or corrugations across spine
Scabrous	Rough to the touch
Smooth	Without surface texture

Shape

Bulbous	Swollen at the base
Curved	Bent from base to tip
Hooked	Apically recurved
Straight	No deviation in long axis
Twisted	Rotated about the long axis

Growth Orientation

Appressed	Parallel to the stem surface contour
Ascending	Angling upward
Deflexed	Abruptly bent downward
Descending	Angling downward
Divergent	Growing at all angles from the areole
Porrect	Perpendicular to the stem surface

Spine Examples

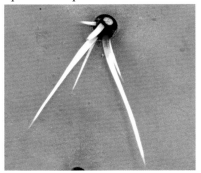

Deflexed spines in bird-foot arrangement of *Opuntia engelmannii*.

Spreading radial spines of *Mammillaria heyderi* var. *meiacantha*.

Porrect central spines are perpendicular to stem surface in *Coryphantha echinus*.

Flattened, ridged, and curved spines of *Echinocactus horizonthalonius*.

Acicular spines of *Opuntia azurea* var. *aureispina*.

Angled and deflexed spines of *Cylindropuntia tunicata*.

Appressed and pectinately arranged radial spines of *Echinocereus reichenbachii*.

How to Use This Book

Descending acicular spines of *Opuntia macrocentra*.

Hooked and twisted spines of *Ferocactus hamatacanthus*.

Hooked central spines along with ascending acicular central and appressed radial spines of *Sclerocactus scheeri*.

Diffuse spines hide the stem of *Echinocereus dasyacanthus*.

Ridged, curved, and flattened descending central spines of *Echinocactus texensis*.

Terete spines with bulbous bases of *Acanthocereus tetragonus*.

Seeds

There are basically two types of seeds found in Texas cacti. The opuntia type is a flattened, rounded, disklike seed. These are about 1–3 mm in width at the longest dimension. The cereus type is shaped like a World War I German helmet and varies in size, shape, color, and surface texture.

Seed Structures

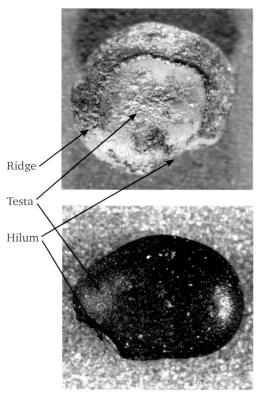

Opuntia-type disk-shaped seed (top) and shiny, black, reticulate cereus type (bottom). Note the cereus type has no ridge.

Hilum:	A scarlike spot where the stalk was attached. This is the basal end of the seed.
Ridge:	A keel-like protective structure on the outer diameter of the seed.
Testa:	The outer seed coat or hard protective coating of the seed. The embryo lies within the testa. It may be varied in color or surface texture.

Seed Characteristics

Cactus seeds have several surface textures and colors. These textures are difficult to discern, requiring a hand lens or microscope to visualize in detail. In this text the following definitions are used:

Discoid:	Disk shaped; a flattened, circular seed
Keeled:	Having a ridge like a boat keel around the seed circumference
Papillate:	Having low, rounded projections
Pitted:	Having small, round depressions
Pyriform:	Pear shaped
Reticulate:	Having mesh- or netlike patterns
Smooth:	Having no discernible surface texture

Cactus Genera of Texas

Genera of Represented Cacti (with number of species described in the text)

Genus	Common Name	No.
Acanthocereus	Triangle cactus	1
Ariocarpus	Living rock cactus	1
Astrophytum	Star cactus	1
Corynopuntia	Dog cholla	4
Coryphantha	Pincushion-beehive cactus	7
Cylindropuntia	Cholla	6
Echinocactus	Barrel cactus	2
Echinocereus	Hedgehog cactus	19
Epithelantha	Button cactus	2
Escobaria	Pincushion cactus	12
Ferocactus	Barrel cactus	3
Lophophora	Peyote	1
Mammillaria	Pincushion cactus	7
Neolloydia	Cone cactus	1
Opuntia	Prickly pear	24
Peniocereus	Night-blooming cereus	1
Pereskia*	Lemonvine	0
Sclerocactus	Fishhook cactus	8
Selenicereus*	Queen of the night	0
Thelocactus	Glory of Texas	3
	TOTAL SPECIES	103

* Purported to remain in extreme South Texas.

Genus *Acanthocereus*: Members of the single Texas species of this genus are long, thin plants frequently unable to support their own weight, dependent upon support from other plants. Supported stems may grow several feet tall. Stems are 1–4 inches in diameter with 3–7 conspicuous ribs. Nocturnal flowers are large and white. These are typically tropical cacti, intolerant of frost and freeze. This species is found near the coast and thrives best on semiarid plains. However, these cacti tolerate moisture well, and when rainfall is adequate, they grow rapidly. Today, *Acanthocereus* has an insecure presence along the coastal habitats of South Texas and Florida. From the Greek *akantha*, "thorn."

Genus *Ariocarpus*: This is a small genus with only one Texas species. The single or cluster of flattened stems measures 2–10 inches in diameter. Frequently, the plants may not rise above the ground surface. Produced from a large carrotlike taproot, the stem is divided into very distinct triangle-shaped tubercles. There are no spines after juvenile growth. This genus flowers in the fall, unusual for Texas cacti. Cacti now classified within *Ariocarpus* have been previously grouped within the genus *Mammillaria*. From the Latin *aria*, "pear," and Greek *karpos*, "fruit."

Genus *Astrophytum*: This genus has only one member in Texas. Plants are usually solitary, globose, brownish green, with no spines, but are covered with white tufts of woolly white hair in the areoles. *Astrophytum* individuals are found only in far South Texas and northern and central Mexico. Cacti now within this genus have been previously classified within the genus *Echinocactus*. From the Greek *astros*, "star," referring to the star-shaped stem in cross section.

Genus *Corynopuntia*: This genus of dog chollas is often regarded as a subgenus of *Opuntia*. Low, creeping, and mat-forming plants have obovoid to cylindrical, jointed stem segments, as well as glochids that are hard to see. Cacti now classified within this genus have previously been listed within the genera *Opuntia* or *Grusonia*.

Genus *Cylindropuntia*: Commonly known as chollas, plants of this genus are often regarded as a subgenus of *Opuntia*. Erect and shrublike, they have round, jointed stems with many segments. They have cylindrical or conical leaves only on young stems, and glochids are present. The joints of these plants may break off easily; the spines becoming lodged in animal and human skin. Cacti now grouped within this genus have been previously classified within the genus *Opuntia*. From the Latin *cylindrus*, "cylinder."

Genus *Coryphantha*: This is one of the most widespread cactus genera, ranging from Canada to Mexico. Plants are globose to cylindrical and often form large clumps. These plants have well-defined tubercles and many spines. They do not have ribs. Moderately large, diurnal flowers open during summer. The fruits are fleshy and may be green or red. Cacti now classified within this genus have been previously classified within the genera *Mammillaria* and *Escobaria*. From the Greek *coryph*, "helmet," and *anthos*, "flower." The generic name means "crest flower," referring to the apical growth of flowers in new tissue at the stem tip.

Genus *Echinocactus*: Known as barrel cacti, these species have numerous rigid spines. The surface of the stem is hard and firm with 8 to more than 20 ribs. Some have densely woolly stem tips and are erect and generally unbranched. Flowers are produced at or near the apex of the plant. Fruits bear scales and sometimes wool but not spines. From the Greek *echinos*, "spiny hedgehog."

Genus *Echinocereus*: One of the largest genera of cacti, these cacti are oval, conical, or cylindrical, always with ribbed stems supporting rows of tubercles. Stems may be branched, forming dense clumps. Spines are mostly straight or slightly curved but never hooked. The flowers are usually large and beautiful, though a few have small, inconspicuous greenish flowers. Flowers are subapical, arising from old growth below the tip of the stem. The fruits are always fleshy and thin skinned and often edible. They are also spiny, but the spines loosen as the fruit matures and may be brushed off. Cacti now classified within this genus have been previously classified within the genus *Cereus*, *Echinocactus*, or *Wilcoxia*. From the Greek *echinos*, "spiny hedgehog," referring to the very thorny covering of this genus; and the Latin *cereus*, "wax candle," derived from the stately appearance of the upright species.

Genus *Epithelantha*: Two species of this button cactus genus grow in the United States, both in Texas. Stems are generally solitary but may occasionally branch, forming clumps rarely exceeding 2 inches in height. One of the smallest cacti, the stem has many tiny tubercles mostly obscured by many minute spines. Spines are so small and depressed that the stem appears smooth to the touch. The tip of the stem is commonly a depression filled with hairlike wool. Flowers arise from new growth at the stem apex. These cacti occur mostly on limestone at lower elevations. Cacti now classified within this genus have been previously classified within the genus *Mammillaria*. From the Greek *epi*, "upon," *thele*, "nipple," and *anthos*, "flower," referring to the flower location near the tubercle apex.

Genus *Escobaria*: Plants are globose to cylindrical, often forming large clumps. The stems may be solitary or branched. They do not have ribs. However, these plants have distinct, well-defined tubercles, and most are well armed with spines. Moderately large flowers open during the day in summer. Flowers are produced in new tissue at the stem apex. This genus name honors Romulo and Numa Escobar, Mexican brothers and botanists. Cacti now listed within this genus have been previously classified within the genera *Mammillaria* and *Coryphantha*.

Genus *Ferocactus*: These barrel cacti are often large, globose to cylindrical, and sometimes branched columnar cacti. Stem tips are not particularly woolly. The cacti often have prominent ribs with conspicuous tubercles producing heavy, long, and often hooked spines. They produce diurnal flowers that arise from new apical growth and bloom in spring and summer. The cacti may grow up to several hundred pounds. The only large barrel cactus in Texas is *Ferocactus wislizeni*, found in the Franklin Mountains near El Paso. Cacti now classified within this genus have previously been classified within the genus *Echinocactus*. From the Latin *ferox*, "wild or fierce," referring to the terrible, "ferocious" spines.

Genus *Lophophora*: Species of this genus are small, globose, or depressed globose cacti that grow from comparatively large, carrotlike taproots. The distinctly blue to blue-green stem grows no more than 2 inches above the ground. The ribs are broad and flat. Stems may be single or may branch from the base to form large clusters. Often glaucous stems have no spines after the early seedling stage. These cacti have a long history of religious and medicinal significance among some Native Americans because of the plant's hallucinogenic alkaloid, mescaline. From the Greek *lophos*, "crest," and *phoreus*, "bearer," referring to the characteristic tufts of hair in the areoles.

Genus *Mammillaria*: This is one of the largest and most popular genera of cacti, yet its members are small and low growing. The stems vary in different species from depressed and almost flat to globular or sometimes columnar. In some species stems remain single, but in others they multiply from the base to form large mounds. In a few species branches may occasionally grow from higher up on the stem. Each stem is entirely covered by tubercles, usually arranged in spiral rows, not forming ribs. Flowers are produced in old-growth tissue from the preceding season, forming a ring or crown around the stem tip. From the Latin *mamilla*, "nipple," in reference to the shape of the tubercles that produce milky white latex in some species.

Genus *Neolloydia*: Members of the single Texas species are small, low-growing, solitary, or loosely clustering plants. The stems are globose or cylindrical and usually with white woolly tips. Ribs are poorly developed or absent. However, conical tubercles are usually prominent. Magenta flowers form on new growth at the tip of the stems. Cacti now classified within this genus have been previously listed within the genus *Mammillaria*, *Echinocactus*, or *Echinomastus*. This cactus is named for Canadian botanist Francis E. Lloyd.

Cactus Genera

Genus *Opuntia*: Covering most of the United States and parts of Canada, this large genus supports the claim that cacti grow over almost the entire United States. Furthermore, in more than half of the states, opuntias are the only cacti found. Characteristics are flat, padlike, jointed stems; cylindrical or conical leaves only on young stems; the presence of glochids; and the production of flowers with areoles that often produce glochids and spines on the ovaries. The erect, pear-shaped fruits are called tunas. This genus, in some sources, includes all the jointed- and segmented-stem cacti, including the chollas (*Cylindropuntia*), dog chollas (*Corynopuntia*), and prickly pears (*Opuntia*). Within this text *Opuntia* includes only the flat-padded, jointed-stem cacti, the prickly pears.

Genus *Peniocereus*: The attractive desert night-blooming cereus has an extremely large, fleshy taproot, from which grow slender stems that are ribbed at first but become round and armed with very short spines. All have fragrant, stunningly beautiful nocturnal flowers produced laterally along the subapical ribs and have rigid spines on the fruits. Texas has only one species. From the Greek *penios*, "thread."

Genus *Pereskia*: These cacti do not look like cacti at all but look like true trees or bushes with woody, branching stems and true photosynthetic leaves. A cactus oddity, members of *Pereskia* lack the large, fleshy stems attributed to common cacti. However, casual inspection reveals that they actually have areoles with large and painful true spines associated with the leaves. The one Texas species doubtfully remains in the Rio Grande Valley. Named for French scholar Nicolas de Peiresc.

Genus *Sclerocactus*: Erect, usually solitary, unbranched plants have globose, ovoid to elongate cylindrical stems with well-defined tubercles. Plants are generally less than 6 inches tall. In some species, strongly hooked central spines, most obvious even on immature plants, are dense, frequently hiding the stem surface. Flowers and fruits arise from new growth near the stem apex. Cacti now classified within this genus have been previously classified within the genera *Ancistrocactus, Echinomastus, Glandulicactus,* and *Toumeya*. From the Greek *sclero*, "hard or cruel," in reference to the formidable spines.

Genus *Selenicereus*: These spindly, vinelike cacti are pencil-thin, stemmed, clambering epiphytic plants that produce large, nocturnal flowers. Typically, stems of this genus are up to 16 feet long and require host plants for support. The existence of the single rare South Texas species is now doubtful. Cacti now classified within this genus have previously been classified within the genus *Cereus*. From the Greek *selene*, "moon."

Genus *Thelocactus*: Known as nipple cacti, plants of this genus may grow singly or in clusters, globose to cylindrical, and usually with distinct tubercles. Flowers are borne on new growth at the tip of the tubercle. Three species occur in Texas. Some cacti now classified within this genus have been previously listed within the genus *Hamatocactus*. From the Greek *thele*, "nipple," referring to the tubercle shape.

SPECIES ACCOUNTS

Corynopuntia aggeria — Clumped dog cholla

Features
Corynopuntia aggeria is a low, many-branched cactus in a mound or a mat as large as 1 m wide. Thick roots are tuberous. Short, clavate to cylindroid stem segments are jointed and strongly persistent, 3.5–9.0 cm long and 1.5–3.0 cm in diameter. New growth occurs by branching from areoles on lower stem segments. Tubercles are prominent, 8–18 mm. Circular, 3–6 mm areoles are covered with dense, white wool and produce spines on the upper two-thirds of the stem segments.

Spines
Central: 3–4 flattened or terete, spreading, 3–5 cm.
Color: Grayish white to reddish brown, with yellow tips and enlarged bases.
Radial: 2–4 slender, terete, deflexed, 6–20 mm.
Color: Grayish white.
Glochids: Numerous, straw to white to rusty brown.

Flowers
Yellow, 4–5 cm. Filaments: yellow-orange to greenish; Anthers: yellow to cream; Stigma: cream-yellow to pale green; Style: pale green to cream.

Fruits
Light yellow, ovoid fruits, aging to gray with tufts of glochids and bristles, 5.5 cm.

Seeds
Cream to brown, subdiscoid seeds, 5–6 mm.

Habitat
This species is found in sandy, silty, or gravelly habitats, often of limestone substrates in southeastern Big Bend in Presidio and Brewster counties at elevations of 580–1,500 m (2,000–5,000 ft). Texas Area 10.

Flowering Season
This dog cholla is expected to bloom in March and April. Flowers open at midday, close in the afternoon, and do not reopen.

Notes
This cactus is one of a group that is difficult to identify. All look similar to *C. schottii*, except this species has a tuberous taproot.

Other Common Name:
Mound-forming opuntia
Synonyms: *Grusonia aggeria, Opuntia aggeria*
Look-alike Species: *C. schottii*

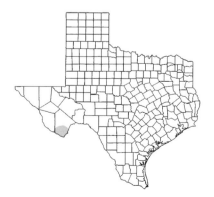

Corynopuntia emoryi   Common devil cholla

Features
Corynopuntia emoryi has larger features than other dog chollas and forms large, sprawling mats up to 6 m across. Fibrous roots produce woody stems that are not easily detached. Stem segments are prostrate or curved upward or erect, clavate and narrowed at the base, 7–19 cm long and 5–50 mm in diameter. New growth originates by branching from lateral areoles. Tubercles are very prominent, 2.5–3.5 cm long. Circular areoles have short, white wool.

Spines
Central: 6–7 flattened, divergent, with bulbous bases, 4.5–7.0 cm.
Color: Tan to yellowish to reddish brown, cross-striated.
Radial: 5–6 slender, terete, deflexed, 1.4–2.5 cm.
Color: Reddish brown or pale yellow.
Glochids: Relatively few.

Flowers
Yellow, 4–7 cm. Filaments: greenish, cream distally; Anthers: cream; Stigma: cream; Style: cream.

Fruits
Light yellow, clavate fruits with mounded wool and glochids, 5–6 cm.

Seeds
Yellowish, smooth, discoid seeds, 4–5 mm.

Habitat
This cholla is found in alluvial substrates of sand or gravel on washes or low hills. Distribution is limited to a small region along the Rio Grande in northwestern Presidio County at elevations of 700–1,200 m (2,300–4,000 ft). Texas Area 10.

Flowering Season
This cactus is expected to bloom from May through late June. Flowers open about midday, close in the afternoon, and do not reopen.

Notes
Other Common Names: Creeping cholla, devil's cholla, Stanley's cholla
Synonyms: *Grusonia emoryi, Opuntia emoryi*
Look-alike Species: *O. schottii*

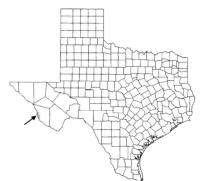

Corynopuntia grahamii — Graham's club cholla

Features
Corynopuntia grahamii forms low mats to 30 cm across with creeping branches, three in a row. Roots are thickened, tuberous taproots. Obovate, cylindrical, and curving stem segments are easily detached, 3–7 cm long and 1.5–3.5 cm in diameter. New growth is formed by branching of apical or upper lateral areoles. Prominent, narrow tubercles are 8–12 mm long. Circular areoles have a covering of dense, white to yellowish wool.

Spines
Central: 4–7 terete, 2.0–3.5 cm.
Color: Gray to straw to brown.
Radial: 6–8 slender, deflexed, 1.0–1.2 cm.
Color: Grayish white.
Glochids: Numerous, 5–7 mm.

Flowers
Yellow, 4–5 cm. Filaments: yellow-orange to greenish; Anthers: yellow to cream; Stigma: cream-yellow to pale green; Style: pale green to cream.

Fruits
Light yellow to gray, ovoid fruits with tufts of white wool, bristles, and glochids, 2.0–3.5 cm.

Seeds
Brown to tan to cream subdiscoid seeds, 5–6 mm.

Habitat
This cholla is found in loose igneous or limestone soils among desert scrub. It is widespread from El Paso southeast to Brewster County at elevations of 700–1,500 m (2,300–5,000 ft). Texas Areas 7–10.

Flowering Season
This cactus is expected to bloom from May to early June. Flowers open about midday for one day only.

Notes
Other Common Names: Graham dog cactus, mounded dwarf cholla
Synonyms: *Grusonia grahamii*, *Opuntia grahamii*
Look-alike Species: *C. schottii*

Corynopuntia schottii — Schott's dog cholla

Features
Corynopuntia schottii is a sprawling cactus that forms extensive, ankle-high mats, 5 m across. Diffuse roots produce green, cylindrical, chain-forming stem segments 3.0–6.5 cm long and 3 cm in diameter. Stems have clavate joints, with new growth branching from lateral areoles. Prominent tubercles have circular areoles with white to dirty white wool.

Spines
Central: 6–8 flattened, divergent, 3.8–7.0 cm.
Color: Reddish brown, dark red, cross-striated.
Radial: 4–6 terete, slender, deflexed, 1.0–1.9 cm.
Color: Grayish white.
Glochids: Relatively few, 3 mm.

Flowers
Yellow, 4–5 cm. Filaments: greenish; Anthers: yellow to cream; Stigma: cream-yellow to pale green; Style: pale green to cream.

Fruits
Light yellow, oval or clavate fruits with narrow tubercles and areoles with wool, glochids, and spines, 4.5 cm.

Seeds
Cream to brown, subdiscoid, pointed seeds, 5–6 mm.

Habitat
This cholla is found in alluvial substrates among Chihuahuan desert scrub and Tamaulipan thorn scrub in two separate ranges from Terrell, Presidio, and Brewster counties to the Pecos River and again in the Rio Grande Valley in Starr, Hidalgo, and Cameron counties at elevations of 100–1,200 m (330–4,000 ft). Texas Areas 2, 6, 7, and 10.

Flowering Season
This cactus is expected to bloom from mid-June to early July. Flowers open about midday, close in the afternoon, and do not reopen.

Notes
Dog chollas are remarkable in that they all open about noon like clockwork.

Other Common Names: Clavellina, dog cactus, Schott's dwarf cholla
Synonyms: *Grusonia schottii*, *Opuntia schottii*
Look-alike Species: *C. aggeria*, *C. grahamii*

Cylindropuntia davisii Davis' cholla

Features
Cylindropuntia davisii is a densely branched shrub, 20–85 cm, stems obscured by spines. Roots have elongate, tuberlike swellings. The plant has a short, woody trunk with exposed wood and vascular gaps clearly visible in older plants. Light green, cylindroid stems are whorled, obscuring the trunks. Stem segments are 4–6 cm long and 8–12 mm in diameter and easily detached. Prominent, laterally compressed, 1–2 cm tubercles have subcircular, 2.0–3.5 mm areoles and tan wool.

Spines
Central: 7–13 flattened, spreading, and obscuring stems, 1.5–5.0 cm.
Color: Yellow to blackish red, tipped yellow covered with loose-fitting, golden sheaths.
Radial: 4–5 acicular, not sheathed, deflexed, 1.0–1.3 cm.
Color: Gray and black speckled.
Glochids: Yellow, 2–4 mm, located at the back of each areole with a tuft of tan wool.

Flowers
Apical on young stem segments. Pale to deep green, to yellow-green, aging to reddish brown, firm, waxy appearance, not opening fully, 4 cm. Filaments: pale green below, purplish above; Anthers: orange-yellow; Stigma: cream to pale purple.

Fruits
Yellow, flattened ovoid, spineless fruit often in chains of two, 2–3 cm.

Seeds
Yellow to tan, subcircular seeds, often sterile, 3 mm.

Habitat
This cholla is found in alluvial soils, especially sandy loam in grasslands and oak-juniper and mesquite woodlands. It is widespread but scattered throughout the western half of Texas at elevations of 600–1,500 m (2,000–5,000 ft). Texas Areas 5–10.

Flowering Season
These cacti are expected to bloom in late May to July. Flowers open mid- to late morning, close at night, and do not reopen.

Notes
Other Common Names: Abrojo, thistle cholla
Synonyms: *Opuntia davisii, O. tunicata* var. *davisii*
Look-alike Species: *C. tunicata*

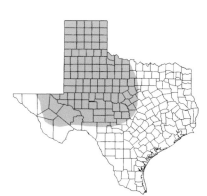

Cylindropuntia imbricata var. arborescens — Tree Cholla

Features
Cylindropuntia imbricata var. *arborescens* is a large, branching shrub or tree with short trunk and rough bark, 1–3 m. It is the largest cactus of West Texas. Gray-green to dark green, cylindrical stem segments, 12–40 cm long and 2–3 cm in diameter. Very prominent tubercles are widely spaced, with elliptic to ovoid areoles with dirty white or gray wool, 5.0–6.5 mm. Patches of nubs of extrafloral nectaries are found at upper edges of areoles.

Spines
Central: 7–8 straight, acicular, spreading, 2–3 cm.
Color: Red-brown or tan, dark reddish brown with silver bases with loose-fitting silver to pale yellow sheaths.
Radial: 3–10 slender, angular, deflexed, 1.2–2.0 cm.
Color: Whitish to gray.
Glochids: Dense axial tuft, pale yellow to brown rimming top of areole, 1–2 mm.

Flowers
Multiple apical flowers on last stem segments are dark pink to purplish to magenta, 5.0–7.5 cm. Colorless forms have been found. Filaments: green to reddish; Anthers: yellow to cream; Stigma: cream; Style: cream to reddish.

Fruits
Yellow, obovoid fruits, 2.4–4.5 cm.

Seeds
Yellow to tan, subcircular seeds, 2.5–4.0 mm.

Habitat
This tree cholla is found in igneous and sedimentary soils of desert grasslands and pinyon-juniper woodlands, widespread and common in the western third of Texas at elevations of 300–1,800 m (1,000–6,000 ft). Texas Areas 2 and 6–10.

Flowering Season
This species is expected to bloom from May through June. Flowers open before midday for one day only.

Notes
Other Common Names: Cane cholla, tree cactus, walking stick cholla
Synonym: *Opuntia imbricata* var. *arborescens*
Look-alike Species: *C. imbricata* var. *argentea* except for size and locality

Cylindropuntia imbricata var. *argentea* — Big Bend cholla

Features
Cylindropuntia imbricata var. *argentea* is a densely branched shrub to 1.2 m with short trunk and grayish appearance due to dense, silvery spines covering the stems. Fibrous roots. Silvery gray-green, cylindroid stem segments, 10–20 cm long and 1.5–4.0 cm in diameter. Silvery green to gray tubercles with closely spaced, elliptic areoles have pale yellow wool, aging to gray.

Spines
Central: 6–14 acicular or slightly flattened, 2–3 cm.
Color: White or pinkish with a greenish base, aging to silver or ivory with pinkish bases. Sheaths silvery white, aging to gray.
Radial: 5–7 slender, deflexed, spreading, 1.8–2.0 cm.
Color: White.
Glochids: Pale yellow at upper edge of areole, 1 mm.

Flowers
Apical, dark magenta flowers, 4.5–5.0 cm. Filaments: reddish purple; Anthers: light to cream yellow; Stigma: cream; Style: reddish purple.

Fruits
Yellow, turbinate fruits, 2.5–3.0 cm.

Seeds
Tan, discoid seeds, 3 mm.

Habitat
This cholla is found on rocky limestone slopes and alluvial flats in desert scrub, with mesquite, lechuguilla, and sotol. Its range is restricted to the very southern parts of Big Bend National Park in extreme southern Brewster County at elevations of 600–700 m (1,900–2,300 ft). Texas Area 10.

Flowering Season
This cactus is expected to bloom from April into July. Flowers open in mid- to late morning, close at night, and do not reopen.

Notes
Other Common Names: Big Bend cane cholla, silver-spine cane cholla
Synonym: Opuntia imbricata var. *argentea*
Look-alike Species: Isolated in its localized habitat; should not be confused

Cylindropuntia kleiniae   Klein cholla

Features
Cylindropuntia kleiniae is a scraggly, sparingly or many-branched shrub with woody trunks, 1.0–2.5 m. It occurs singly or in thickets. Green to gray-green cylindrical stem segments are easily detached, 4–25 cm long and 6–12 mm in diameter. Oval or round areoles are surrounded by purplish pigment, with dense, short gray and longer yellow mounded wool.

Spines
Central: 1–4 acicular, straight to slightly arched, 1–3 cm.
Color: Yellow to gray with translucent tip, loosely sheathed with golden or silver sheaths.
Radial: 0–6 slender, bristly, 0.5–4.0 mm.
Color: Pale yellow or gray.
Glochids: Yellow, in tufts at the edge of the areoles, 2–3 mm.

Flowers
Pinkish to magenta or purple-tinged flowers or pale greenish, greenish cream, or maroon-tinged, 2.5–5.0 cm. Filaments: greenish to pinkish red; Anthers: yellow; Stigma: cream to pinkish; Style: green to pinkish maroon.

Fruits
Green to red or orange, obovoid fruits, 1.3–3.4 cm.

Seeds
Tan, oval to squarish seeds, 3.5–5.0 mm.

Habitat
Klein cholla occurs in alluvial flats of sandy loam or rocky slopes of limestone and grows along with creosotebush and mesquite. Its range is generally the western half of the Trans-Pecos with the greatest concentration near the Rio Grande in Big Bend at elevations of 500–1,500 m (1,600–5,000 ft). Texas Area 10.

Flowering Season
This cholla is expected to bloom from late May through August. Flowers open mid- to late morning, close at night, and do not reopen.

Notes
Other Common Names: Candle cholla, Klein's pencil cholla
Synonyms: Opuntia kleiniae, O. wrightii
Look-alike Species: C. leptocaulis

Cylindropuntia leptocaulis Christmas cholla

Features
Cylindropuntia leptocaulis is a compact, low shrub with intricate, alternate pencil-thin branched stems. Plants grow to 1 m or more and may be solitary, entwined among other vegetation or in fences. Diffuse roots produce yellow-green to dark green or purplish and cylindrical stem segments, 2–40 cm long and 8–9 mm in diameter. Linear tubercles have broadly elliptic, 2–3 mm areoles with short gray and longer yellow wool, aging to gray. Two growth forms, "long-spine" and "short-spine," are found in Texas.

Spines
Central: "Short-spine" form: 1–3 deflexed or porrect, acicular or somewhat flattened, 4–30 mm. "Long-spine" form: 1–3 strongly porrect, loosely sheathed, to 4.8 cm.
Color: Gray, tipped with yellow; when present, sheaths are golden or silver, tipped with gold.
Radial: 1–7 bristlelike spines in young areoles, 1–3 mm.
Color: As the centrals.
Glochids: Yellowish to reddish brown in tufts at areole margins, 1–3 mm.

Flowers
Flowers arise on distal halves of larger joints and are greenish yellow to yellow, 1.0–2.2 cm. Filaments: pale green to greenish yellow; Anthers: pale to creamy yellow; Stigma: cream to white.

Fruits
Yellow to red, obovoid fruits, 9–15 mm.

Seeds
Pale yellow, smooth, oval to squarish seeds, 3.0–4.5 mm.

Habitat
Christmas cholla grows in sandy, loamy, to gravelly soils in desert grasslands and oak-juniper woodlands throughout most of Texas at elevations of 0–2,500 m (0–8,000 ft). Texas Areas 2–10.

Flowering Season
This cactus is expected to bloom from March through August and often later. Flowers open late afternoon, close at night, and do not reopen.

Notes
Other Common Names: Christmas cactus, pencil cactus, tasajillo
Synonym: *Opuntia leptocaulis*
Look-alike Species: Juvenile *C. kleiniae*

Cylindropuntia tunicata Icicle cholla

Features
Cylindropuntia tunicata is a compact, densely branched shrub forming low, spreading clumps 1 m in diameter and to 35 cm tall. Plants have sheathed spines that glisten and partially obscure the stems. Diffuse roots produce thick, woody stems with pale green to green stems, segments 5–20 cm long and 1.5–2.5 cm in diameter. Segments easily detached. Very prominent, 2–3 cm tubercles have oval, 3–7 mm areoles. Extrafloral nectaries appear as 2 or 3 short, peg-like structures on upper side of young areoles, emerging slightly above felt-like, dirty white to yellow wool.

Spines
Central: 1–5 flattened, sheathed, 2.5–5.0 cm.
Color: Pale yellowish tan, drying to reddish brown, sheaths silvery white.
Radial: 2–4 slender, terete, deflexed, 5–15 mm.
Color: Gray to tan.
Glochids: In tuft at upper edge of areole, pale yellow, 0.5–1.2 mm.

Flowers
Apical on most distal stem segments, yellow to greenish yellow, 3.5–5.0 cm. Filaments: greenish; Anthers: yellow; Stigma: greenish to greenish yellow; Style: greenish.

Fruits
Yellow to yellow-green, obconic fruits, usually sterile, 2.5–5.0 cm.

Seeds
Light tan, obovate, smooth seeds, 2.5 mm.

Habitat
This cholla grows on sandy, gravelly, calcareous slopes of desert grasslands among grasses and other desert shrubs. Its range is limited to small populations in Brewster and Pecos counties at elevations of 1,500 m (5,000 ft). Texas Area 10.

Flowering Season
This cactus is expected to bloom from May through June and perhaps later. Plants may not bloom every year.

Notes
Other Common Names: Clavellina, sheathed cholla, thistle cholla
Synonyms: *Opuntia stapeliae, O. tunicata*
Look-alike Species: None

Opuntia atrispina — Black-and-yellow-spined prickly pear

Features
Opuntia atrispina is an erect or spreading shrub, 0.5–1.0 m tall. It is identified by black central spines with yellow distal portions. Green, obovate to circular pads are 10–17 cm long and 7.5–15.0 cm wide. Areoles are 3–5 mm, ovate to oblong and 5–7 per diagonal row across the midstem segment with tan wool. Spines are mostly in distal areoles on upper half of pad.

Spines
Central: 4–7 acicular, porrect or slightly descending, 2.5–3.5 cm.
Color: Black to reddish brown, tipped yellow.
Radial: 1–2 straight, acicular, conspicuous, 1 cm.
Color: Brown to yellow tips, fading to gray.
Glochids: Profuse and small in bushy crescent, yellow to brown, aging blackish, to 5 mm.

Flowers
Yellow throughout or with greenish centers fading to apricot, 2.5–6.5 cm. Filaments: cream; Anthers: cream; Stigma: cream to pale green; Style: white to rose.

Fruits
Red to purple, spherical or obovoid, spineless fruits, 1.5–2.5 cm. May have 20–25 areoles.

Seeds
Tan to gray, subcircular seeds, 3–4 mm.

Habitat
This prickly pear is found in hills of limestone substrates in desert grasslands and in scrublands in the southeastern Trans-Pecos to Uvalde County at elevations of 400–700 m (1,300–2,300 ft). Texas Areas 6, 7 and 10.

Flowering Season
This cactus is expected to bloom in April and May. Flowers open for 2–3 days.

Notes
Other Common Name: Dark spined opuntia
Synonyms: None
Look-alike Species: *O. strigil*

Opuntia aureispina — Golden-spined prickly pear

Features
Opuntia aureispina is a treelike shrub with several ascending stems from short spiny trunks up to 1.5 m high. Yellow spines appear more numerous and longest on upper part of pads. Glaucous, light blue-green to yellow-green, circular to broadly obovate pads 8–16 cm long and 8–12 cm wide. Areoles are 4–5 mm, ovate to oblong, and 6–8 per diagonal row across the midstem segment, with brown to black wool. Spines are found in areoles from upper margin to base of pad.

Spines
Central: 3–5 acicular, 2–6 cm.
Color: Bright yellow-orange to light brown or nearly black, tipped yellow.
Radial: 1–7 slender, deflexed, 2 cm.
Color: Reddish brown at base.
Glochids: Yellow, spaced in narrow row circling areoles, to 5 mm.

Flowers
Yellow with orange or red centers, 5–7 cm. Filaments: yellow to pale green; Anthers: pale yellow; Stigma: pale green or pale yellow; Style: yellowish or pink.

Fruits
Pale green to reddish tan, ovoid fruits with several yellow spines, 3–4 cm.

Seeds
Tan to light brown, irregularly shaped seeds, 3–6 mm.

Habitat
This cactus is found in fractured limestone of low hills and desert flats near Mariscal Mountain and Boquillas Canyon in Big Bend in Brewster County at elevations of 500–600 m (1,500–1,900 ft). Texas Area 10.

Flowering Season
This species is expected to bloom in late March and April. Flowers open mid- to late morning, close at night, and usually do not reopen.

Notes
Other Common Names: None
Synonyms: *O. azurea* var. *aureispina*, *O. macrocentra* var. *aureispina*
Look-alike Species: None

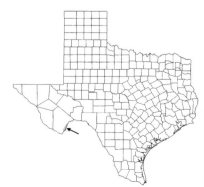

Opuntia azurea var. *diplopurpurea* — Long-spined purplish prickly pear

Features
Opuntia azurea var. *diplopurpurea* is a low, sprawling shrub, to 30–60 cm, without a trunk. It is one of a few prickly pears with purple stems. Pads are blue-gray, blue-green, or purplish with purple near areoles and pad margins. Pads are obovate to subcircular, 14–20 cm long and 14–20 cm wide. Areoles are elliptic to circular with 4–5 in diagonal rows on midstem segment. Spines are found on upper half of pad or sometimes only on upper margin.

Spines
Central: 2–6 acicular, one strongly deflexed, 5–12 cm.
Color: Black to dark reddish brown with white tips.
Glochids: Reddish brown in dense tufts at areole margin, 1–6 mm.

Flowers
Yellow flowers with red centers, 2.5–3.5 cm. Filaments: pale green to cream; Anthers: yellowish; Stigma: cream to pale green; Style: cream.

Fruits
Red to purplish, ovate to obovate fruits, 1–3 cm.

Seeds
Tan, subcircular, flat seeds, 3.8–4.5 mm.

Habitat
This prickly pear is commonly found on desert flats and mountain grasslands from Hudspeth County southeastward into Brewster County in the Trans-Pecos at elevations of 750–1,650 m (2,500–5,500 ft). Texas Area 10.

Flowering Season
This cactus is expected to bloom from April through June. Flowers open in midmorning, close at night, and usually do not reopen unless opening delayed by clouds.

Notes
Other Common Names: Long-spine prickly pear, purple prickly pear
Synonyms: *O. azurea, O. violacea* var. *macrocentra*
Look-alike Species: *O. macrocentra*

Opuntia azurea var. *discolor*   Big Hill prickly pear

Features
Opuntia azurea var. *discolor* is an upright, spreading shrub, to 1 m. Pads are blue-green with a purplish tinge near the areoles, broadly obovate, 20 cm long and 20 cm wide. Areoles are elliptic with 5–6 in diagonal rows on midstem segment. Spines are found on upper third or half of pad.

Spines
Central: 1–4 acicular, projecting, curving, and twisting, 3–10 cm.
Color: Yellow with dark brown mottling to black.
Glochids: Yellow to brown in crescent at upper areole margin, 1–6 mm.

Flowers
Yellow flowers with red basal portions and bright red centers, 5–8 cm. Filaments: green; Anthers: yellow; Stigma: cream to pale green; Style: cream.

Fruits
Red to purplish, ovate to obovate fruits, 1–3 cm.

Seeds
Tan, subcircular, flat seeds, 3.8–4.5 mm.

Habitat
This prickly pear is found on slopes of igneous hillsides along the Rio Grande in far southern Presidio County at elevations of 750–900 m (2,500 to 3,000 ft). Texas Area 10.

Flowering Season
This cactus is expected to bloom in April and May.

Notes
Other Common Names: None
Synonyms: None
Look-alike Species: *O. camanchica*

Opuntia azurea var. *parva* Long-spined purplish prickly pear

Features
Opuntia azurea var. *parva* is a compact, sprawling or upright, spreading shrub, to 1 m, sometimes with a short trunk. It is one of a few prickly pears with purple stems. Pads are glaucous blue-gray, blue-green, or purplish with a purple tinge around areoles and pad margins. Smallish pads are obovate, 6–19 cm long and 6–14 cm wide. Areoles are elliptic in 3–5 diagonal rows on midstem segment with tan to white wool, aging to black. Spines are found on upper half of pad or only on upper margin.

Spines
Central: 1–3 acicular, one strongly deflexed, 5–12 cm.
Color: Black to dark reddish brown with white tips.
Glochids: Yellow to reddish brown in dense crescent at areole margin, 1–6 mm.

Flowers
Yellow flowers with red basal portions and bright red centers, 5–8 cm. Filaments: pale green to cream; Anthers: yellow; Stigma: cream to pale green; Style: cream.

Fruits
Red to purple, ovate to obovate fruits with deep umbilicus, 1.5–3.0 cm.

Seeds
Tan, discoid seeds, 5.0–5.7 mm.

Habitat
This is one of the most common prickly pears. It is found on the desert floor around the Chisos Mountains in Big Bend National Park and northward in Brewster County at elevations of 575–1,100 m (1,900–3,700 ft). Texas Area 10.

Flowering Season
This cactus is expected to bloom from April through May. Flowers open in midmorning, close at night, and do not reopen unless delayed by clouds.

Notes
Other Common Names: Long-spine prickly pear, purple prickly pear
Synonyms: *O. azurea, O. violacea* var. *macrocentra*
Look-alike Species: None

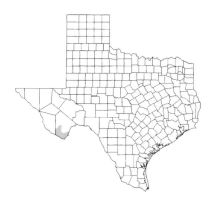
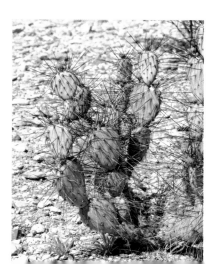

Opuntia camanchica — Comanche prickly pear

Features
Opuntia camanchica is said to be a part of the *O. phaeacantha* complex of several brown-spined prickly pear varieties. It is recognized herein as a low, spreading to erect shrub. During the winter months it seems to be even more erect. Erect plants are less than 70 cm tall. Pads are green, sometimes reddish during stress, obovate to subcircular, 15–25 cm long and 15–20 cm wide. Areoles are oval, 5–8 mm with gray wool.

Spines
Central: 1–2, lower acicular or flattened and deflexed, upper deflexed or projecting, 1–3 cm.
Color: Whitish to gray with yellow, red, or brown tips.
Radial: When present (rarely), 2 slender, acicular, 3–6 mm.
Color: White to gray or reddish brown.
Glochids: Yellowish brown in crescent around areoles, sparse on lower areoles, 1–3 mm.

Flowers
Yellow with pale or dark red centers, 5–7 cm. Filaments: cream; Anthers: yellow; Stigma: cream to dark green; Style: cream.

Fruits
Red, obovate fruits, with few spines near upper edge, 3.5–5.0 cm.

Seeds
Tan to light brown discoid seeds, 4–5 mm.

Habitat
This species is found in sandy to rocky soils of deserts and semidesert lands throughout most of the west half of Texas, including the Panhandle at elevations of 650–1,500 m (2,000–4,600 ft). Texas Areas 5–10.

Flowering Season
This prickly pear is expected to bloom from April through mid-June. Flowers open midmorning, close at night, and seldom reopen.

Notes
Other Common Name: Camanchican prickly pear
Synonyms: *O. phaeacantha* var. *camanchica*, *O. phaeacantha* var. *major*
Look-alike Species: *O. phaeacantha* varieties

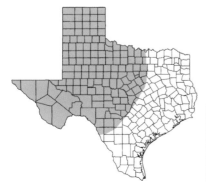

Opuntia chisosensis — Chisos prickly pear

Features
Opuntia chisosensis is an erect shrub, to 1 m. This medium-sized prickly pear branches from fibrous roots and a thick base, not a trunk, with 6 pads per branch and conspicuous bright yellow spines. Bluish, gray-green, smooth and glabrous stems are circular to broadly obovate, 15–30 cm long and 12–25 cm wide. Areoles are 3–8 mm, elliptic to obovate and 5–7 per diagonal row across the midstem segment, with tan wool, aging to black.

Spines
Central: 1–5, flattened or one acicular, deflexed, 2–6 cm.
Color: Bright yellow to reddish orange to nearly black with pale yellow tips.
Radial: 0–1 acicular, often deflexed, 3–6 cm.
Color: As the centrals.
Glochids: Yellowish, in widely spaced crescent at areole margin, 2–4 mm.

Flowers
Pale yellow flowers, fading to salmon, 5.0–6.5 cm. Filaments: pale green; Anthers: yellow; Stigma: green; Style: yellow.

Fruits
Red or reddish purple, ellipsoid to spherical fruits with 16–20 areoles, spineless, 3.3–4.5 cm.

Seeds
Yellow to tan, irregularly shaped, flattened seeds, 3.5–4.5 mm.

Habitat
This uncommon prickly pear is found in grasslands and oak-juniper-pinyon woodlands in the Chisos Basin mountains. Found in Brewster County at elevations of 600–2,200 m (5,200–7,200 ft). Texas Area 10.

Flowering Season
This species is expected to bloom from May through June. Flowers open late morning, close in the evening, and may not reopen.

Notes
Other Common Names: None
Synonym: *O. lindheimeri* var. *chisosensis*
Look-alike Species: *O. engelmannii*

Opuntia chlorotica — Clockface prickly pear

Features
Opuntia chlorotica is an erect, spreading shrub, to 2.5 m, unique in prickly pears in that spines cover the short, well-defined trunk. Pads are yellowish green, nearly smooth, and glabrous. Pads are circular to broadly obovate, 13–21 cm long and 11.5–19.0 cm wide. Areoles are elliptic and cover the pads and trunk. Spines are found on upper half of pad or only on upper margin.

Spines
Central: 1–7 acicular or somewhat flattened, deflexed or curving, 1.2–5.0 cm.
Color: Pale yellow.
Glochids: Yellow in dense crescent at areole margin, large, up to 1.4 cm.

Flowers
Yellow flowers, 3–6 cm. Filaments: whitish yellow; Anthers: white; Stigma: white or yellow; Style: yellow.

Fruits
Reddish purple, spherical to ovate fruits, 3–6 cm.

Seeds
Yellowish, flat, reniform seeds, 3.5 mm.

Habitat
This prickly pear is found as an accidental species in Texas in Tamaulipan thorn scrub in Starr and Zapata counties. Elsewhere in New Mexico and westward it is found in chaparral and desert grasslands at elevations of 600–2,400 m (1,900–7,900 ft). Texas Area 6.

Flowering Season
This cactus is expected to bloom from April through July.

Notes
Other Common Names: Flapjack prickly pear, pancake prickly pear
Synonym: *O. palmeri*
Look-alike Species: *O. engelmannii* var. *lindheimeri*

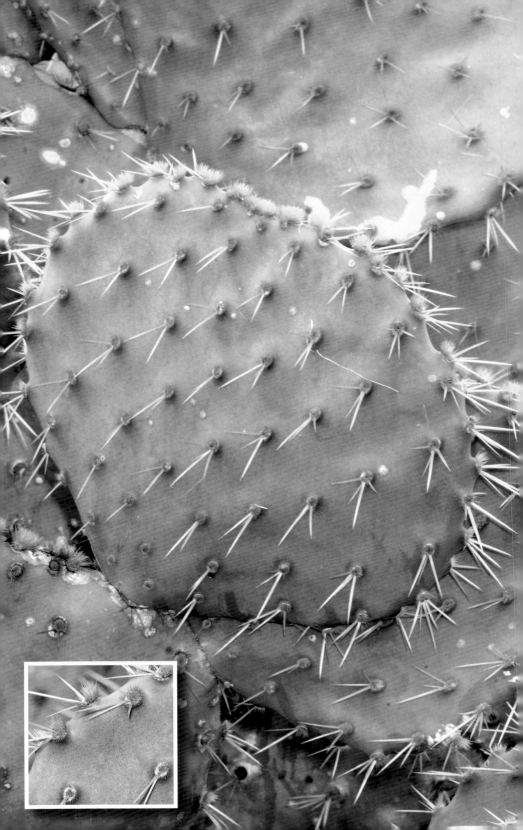

Opuntia engelmannii var. *alta* — Tall prickly pear

Features
Opuntia engelmannii var. *alta* is a relatively large, sprawling shrub, 0.5–3.0 m tall and as wide as 5 m. Erect or ascending branches have large, heavy pads with mostly yellow spines. Green to blue-green, broadly obovate to subcircular pads, 15–30 cm long and 12.5–25.0 cm wide, growing from a fairly definite trunk. Areoles are elliptic to almost circular, 4.5–6.0 mm, and filled with tan to gray wool.

Spines
Central: 1–12 acicular or flattened, deflexed, 2.5–6.5 cm.
Color: Translucent yellow.
Radial: 0–2 variable, acicular, 1–3 cm.
Color: Mostly yellow.
Glochids: Around periphery of areole, yellow, 3–6 mm.

Flowers
Variable from cream through yellow, orange, and red, 5.0–7.5 cm. Filaments: cream; Anthers: cream; Stigma: pale green; Style: greenish yellow.

Fruits
Reddish purple to purple, ovate fruits with numerous glochids and tiny spines, 3–7 cm.

Seeds
Tan, irregularly discoid seeds, 1.5–2.5 mm.

Habitat
This prickly pear is found from the mouth of the Rio Grande northward to the Louisiana border within 20 miles of the Gulf Coast. It grows in thickets around the mouths of rivers and more sparsely behind the beaches and dunes at elevations of 0–100 m (0–350 ft). Texas Areas 1, 2, and 6.

Flowering Season
This cactus is expected to bloom from March through June. Flowers open in mornings and afternoons, may close at night, and may reopen a second day.

Notes
This species is one of several varieties of prickly pear and has undergone many name changes over the years.
Other Common Names: Nopal, tuna
Synonym: *O. lindheimeri*
Look-alike Species: *O. engelmannii* var. *lindheimeri*

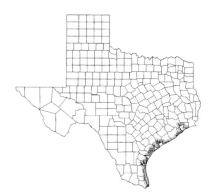

Opuntia engelmannii var. *engelmannii* — Engelmann's prickly pear

Features
Opuntia engelmannii var. *engelmannii* is the most widespread and visible prickly pear in far West Texas. It is recognized as a shrub or tree with a short trunk and upright or spreading branches to 1.0–1.5 m tall. Pads are pale green, circular to obovate, 20–30 cm long and 5–25 cm wide. The stems branch many times. Pads are not easily detached. Areoles are 4–7 mm, subcircular, and 5–8 per diagonal row across the midstem segment, with tawny wool, aging to black. Spines are generally white.

Spines
Central: 3–4 acicular, deflexed in a "bird-foot" pattern, 1–4 cm.
Color: Bone to chalky white.
Radial: 1–2 acicular, 3.5–10.0 mm.
Color: Yellow or white.
Glochids: Yellow, conspicuous and stout, of unequal lengths and widely spaced around areole periphery.

Flowers
Widely funnelform, clear yellow, 7–8 cm. Filaments: cream to pale green; Anthers: yellow; Stigma: green; Style: cream.

Fruits
Purple to dark beet red, barrel-shaped or obovate fruits, 5.5–8.0 cm.

Seeds
Tan, irregular discoid seeds, 3–4 mm.

Habitat
This prickly pear is found in sandy to rocky habitat of desert to mountain grasslands and woodlands. It is common, with a range widespread through the Trans-Pecos at elevations of 300–2,700 m (1,000–8,400 ft). Texas Areas 7–10.

Flowering Season
This species is expected to bloom from April through July, or later dependent upon altitude. Flowers open in morning or by midday, close at night, and do not reopen.

Notes
Other Common Names: Discus cactus, nopal, purple-fruited prickly pear, tulip prickly pear
Synonyms: *O. discata, O. engelmannii, O. engelmannii* var. *discata, O. lindheimeri, O. phaeacantha* var. *discata*
Look-alike Species: *O. engelmannii* var. *lindheimeri*

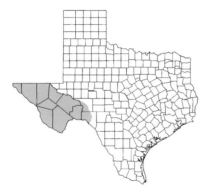

Opuntia engelmannii var. *lindheimeri*   Texas prickly pear

Features
Opuntia engelmannii var. *lindheimeri* is a relatively large, sprawling shrub, 0.5–3.0 m tall and as wide as 5 m. Erect or ascending branches have large, heavy pads with mostly yellow spines, not arranged in a "bird-foot" pattern. Green to blue-green, broadly obovate to subcircular pads, 15–30 cm long and 12.5–25.0 cm wide. Areoles are elliptic to almost circular, 4.5–6.0 mm, and filled with tan to gray wool.

Spines
Central: 1–8 acicular, deflexed, 1.5–7.0 cm.
Color: Clear yellow, sometimes whitish yellow, often with reddish brown bases.
Radial: 0–2 variable, acicular, deflexed, 1–3 cm.
Color: Mostly yellow.
Glochids: Around periphery of areole, dirty yellow, 3–6 mm.

Flowers
Brilliant yellow without red centers, 5.0–7.5 cm. Filaments: cream; Anthers: cream; Stigma: dark green; Style: greenish yellow.

Fruits
Reddish purple to purple, pyriform, spineless fruits, 3–7 cm.

Seeds
Tan, irregularly discoid seeds, 3–4 mm.

Habitat
Widespread, Texas prickly pear is found in deep, sandy to gravelly or rocky soils within open plains to woodlands and chaparral, in Chihuahuan desert scrub, and in Tamaulipan thorn scrub. More common east of the Pecos River to the Edwards Plateau and into South Texas at elevations of 0–1,900 m (0–6,200 ft). Texas Areas 1–8 and 10.

Flowering Season
This cactus is expected to bloom from March through June. Flowers open in mornings and afternoons, may close at night, and may reopen a second day.

Notes
This species is one of several varieties of prickly pear and has undergone many name changes over the years. Collectively, the prickly pear is the State Plant of Texas.
Other Common Names: Lindheimer's prickly pear, nopal, tuna
Synonym: *O. lindheimeri*
Look-alike Species: *O. engelmannii* var. *engelmannii*

Opuntia engelmannii var. *linguiformis*   Cow-tongue prickly pear

Features
Opuntia engelmannii var. *linguiformis* is a relatively large, sprawling shrub, 1–2 m tall. Erect or ascending branches have long, green to blue-green, "tongue-shaped" pads to 100 cm long and 10–15 cm wide. Areoles are elliptic to almost circular, 4.5–6.0 mm, and filled with tan to gray wool. This species is now accepted as a cultivated species where found, presumably extinct in the wild.

Spines
Central: 1–3 acicular, deflexed, 1–2 cm.
Color: Yellow to reddish brown.
Radial: 0–2 variable, acicular, deflexed, 1–3 cm.
Color: Mostly yellow.
Glochids: Around periphery of areole, dirty yellow, 3–6 mm.

Flowers
Yellow to yellowish orange, or uncommonly, orange or red, 5.0–7.5 cm. Filaments: yellow; Anthers: yellow; Stigma: green; Style: greenish yellow.

Fruits
Purplish, pyriform, spineless fruits, 3–7 cm.

Seeds
Tan, irregularly discoid seeds, 3–4 mm.

Habitat
This species is now found commonly cultivated as an ornamental in Central and West Texas at elevations of 100–1,600 m (330–5,250 ft). Texas Areas 5–7 and 10.

Flowering Season
This cactus is expected to bloom from April through June. Flowers open mornings and afternoons for only one day.

Notes
This species, first discovered in the wild near San Antonio, is now found cultivated throughout the state. Many landscapers use this cactus for its unusual-looking pads.
Other Common Name: Lenqua de vaca
Synonyms: *O. lindheimeri* var. *linguiformis*, *O. linguiformis*
Look-alike Species: None

Opuntia humifusa   Eastern prickly pear

Features
Opuntia humifusa has the appearance of a low, prostrate, clump-forming shrub, usually only 1 or 2 stem segments tall, growing to 30 cm high. Roots are fibrous, uncommonly tuber-like. Pads are dark or bright, shiny green, obovoid to circular, wrinkled under stress, glabrous and fleshy or flabby, 5.0–17.5 cm long and 4–6 cm wide. Areoles are 2–4 mm, oval to circular, with 4–6 per diagonal row on midstem segment with tan to brown wool. Spines on upper areoles only.

Spines
Central: 0–2 straight, terete, spreading, 1 porrect, 2.5–6.0 cm. Often absent.
Color: Whitish to brownish.
Glochids: Yellow to reddish brown in dense adaxial crescent, 3 mm.

Flowers
Pale to bright yellow with red centers or streaked with red, 2–3 cm. Filaments: orange; Anthers: pale yellow to cream; Stigma: pale greenish yellow.

Fruits
Greenish to reddish fruits, tapering at base with 10–18 areoles and glochids, 3–5 cm.

Seeds
Tan, thick, subcircular seeds, 3.5–4.5 mm.

Habitat
Eastern prickly pear is found in sandy soils or dry woodlands and hillsides. Its range is large, extending from Central and East Texas north to South Dakota, and east to Florida and New York, at elevations of 0–1,000 m (0–3,200 ft). Texas Areas 1–7.

Flowering Season
This species is expected to bloom from February through August.

Notes
Other Common Names: Low prickly pear, smooth prickly pear
Synonym: *O. compressa* var. *humifusa*
Look-alike Species: *O. macrorhiza*

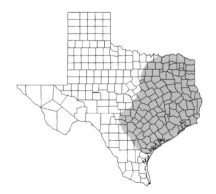

Opuntia macrocentra — Long-spined purplish prickly pear

Features
Opuntia macrocentra is an erect, spreading shrub, to 1 m, rarely with a trunk. It is one of a few prickly pears with purple stems. Stems are blue-gray, blue-green, or purplish with purple near areoles and pad margins. Pads are obovate to subcircular, 7–20 cm long and 6–20 cm wide. Areoles are elliptic to circular with 6–8 in diagonal rows on midstem segment, with tan to white wool, aging to black. Spines are found on upper half of pad or only on upper margin.

Spines
Central: 1–4 acicular, one strongly deflexed, 5–12 cm.
Color: Black to dark reddish brown with white or yellowish tips.
Glochids: Yellow to reddish brown in dense crescent at areole margin, 1–6 mm.

Flowers
Yellow flowers with red basal portions and bright red centers, 5.5–8.0 cm. Filaments: pale green to cream; Anthers: yellowish; Stigma: cream to pale green; Style: cream.

Fruits
Red to purplish, obovoid to barrel-shaped fruits, 2.5–4.0 cm.

Seeds
Yellowish to tan, subcircular to reniform seeds, 5–7 mm.

Habitat
This prickly pear is found on sandy desert flats, rocky hills, and valley grasslands from El Paso County eastward into Reeves County in the Trans-Pecos at elevations of 900–1,600 m (2,800–5,000 ft). Texas Areas 6, 7, and 10.

Flowering Season
This cactus is expected to bloom from April through May. Flowers open in midmorning, close at night, and do not reopen.

Notes
Other Common Names: Long-spine prickly pear, purple prickly pear, red-eye prickly pear
Synonyms: *O. violacea*, *O. violacea* var. *macrocentra*
Look-alike Species: None

Opuntia macrorhiza Plains prickly pear

Features
Opuntia macrorhiza is a low, creeping, clump-forming shrub, 30–40 cm tall. Roots are somewhat tuberlike. Stems are wrinkled when stressed. Dark dull green, obovate to circular, flattened pads, 5–13 cm long and 8–12 cm wide. Areoles are oval to subcircular, quite tuberculate, 2–4 mm, with dense brownish to gray wool. Spines are found on upper part of pads.

Spines
Central: 1–3, one lower flattened and deflexed, others terete, projecting, 3.0–5.5 mm.
Color: Whitish.
Radial: 1–2, slender, deflexed when present.
Color: As the centrals.
Glochids: Pale yellow or reddish brown in dense tuft, 2–4 mm.

Flowers
Yellow flowers with red centers, 5–8 cm. Filaments: pale yellow; Anthers: yellow; Stigma: cream; Style: cream.

Fruits
Green to yellowish to red, elongated obovoid fruits, 2.5–4.0 cm.

Seeds
Tan, subcircular seeds, 4–5 mm.

Habitat
This cactus is found in deep sand, loam, and limestone habitat on prairies and in chaparral, grassy woodlands, and conifer forests. Its range is widespread from about Houston, northward into Oklahoma, New Mexico, and beyond at elevations of 100–2,300 m (330–7,200 ft). Texas Areas 1–9.

Flowering Season
This prickly pear is expected to bloom from April through June. Flowers open for one day only.

Notes
Other Common Names: Chain prickly pear, grassland prickly pear, western prickly pear
Synonym: *O. compressa* var. *macrorhiza*
Look-alike Species: None

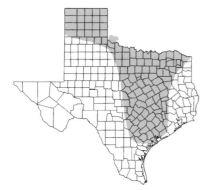

Opuntia phaeacantha — Brown-spined prickly pear

Features
Opuntia phaeacantha is a complex of several medium-sized, brown-spined prickly pear varieties. These are recognized herein as somewhat erect shrubs with sprawling, weak stems and pads that sag or lie on the ground, especially during the winter months. Stems are decumbent to commonly trailing to 0.3–1.0 m. Pads are green to dark green, sometimes reddish during stress, obovate to elliptic, 10–25 cm long and 9–18 cm wide. Areoles are 3–6 mm, oval to elliptic, with 5–7 per diagonal row on midstem segments, with tan to brown wool, aging to gray. Spines are on upper fourth of stem or essentially absent.

Spines
Central: 0–3, flattened to terete, twisted, deflexed, 3–8 cm.
Color: Whitish to tan to brown, reddish brown at bases.
Radial: 2 slender, acicular, 4.5–8.0 mm.
Color: Off white to yellowish.
Glochids: Yellowish or yellow-brown, dense tuft, to 3 mm.

Flowers
Yellow with pale or dark red centers, 5–7 cm. Filaments: greenish to pale yellow; Anthers: yellow; Stigma: yellowish, pale or dark green; Style: cream to pinkish.

Fruits
Wine red to purple, obovate to barrel-shaped fruits, 3–5 cm with 18–24 spineless areoles.

Seeds
Tan, subcircular, notched seeds, 4–5 mm.

Habitat
This species is found in sandy to rocky soils of deserts, chaparral, and surrounding mountains and plains. Its distribution is from the Trans-Pecos and Big Bend eastward through the Edwards Plateau and north into the Panhandle at elevations of 200–2,100 m (600–6,500 ft). Texas Areas 5–10.

Flowering Season
This prickly pear is expected to bloom from May through June. Flowers open midday or early afternoon, close at night, and seldom reopen.

Notes
Other Common Names: New Mexico prickly pear, purple-fruited prickly pear, tulip prickly pear
Synonyms: *O. phaeacantha* var. *major*, *O. phaeacantha* var. *phaeacantha*
Look-alike Species: *O. camanchica*, *O. tortispina*

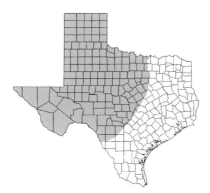

Opuntia polyacantha var. *trichophora*   Southern Plains prickly pear

Features
Opuntia polyacantha var. *trichophora* is a small, low, prostrate shrub, 10–25 cm high, with small pads partially obscured by spines. Roots are produced from stems where they may touch the ground. Green, broadly ovate to circular pads, 7–13 cm long and 5.5–11.0 cm wide. Areoles are oblong to circular, 2.0–3.5 mm. Spines present in all but the lowest areoles.

Spines
Central: 4–7 acicular, ascending and descending, 4–7 cm.
Color: Whitish, pale reddish tips or gray to pale reddish overall.
Radial: 4–6 acicular, appressed, 3–6 mm.
Color: Whitish with pale yellow or reddish tips.
Glochids: Light yellow, tightly packed in apical tuft in areole, 1.5–5.0 mm.

Flowers
Yellow flowers, 4–5 cm. Filaments: white, green, or pale yellow; Anthers: cream yellow; Stigma: dark green; Style: cream to pinkish.

Fruits
Spiny, green to dull red, obconic fruits, 1.9–2.5 cm, with 12–42 cottony areoles and 6–15 spines.

Seeds
Very large, tan or cream, irregularly discoid, flattened seeds, 6.0–6.8 mm.

Habitat
This cactus is found in clay, sand, or gravel in grasslands, scrubland, and pinyon-juniper woodlands within two disjointed areas of the Trans-Pecos and the Panhandle at elevations of 500–2,000 m (1,500–6,200 ft). Texas Areas 8–10.

Flowering Season
This species is expected to bloom from May through June. Flowers open midday, close at night, and infrequently reopen.

Notes
Other Common Names: Plains prickly pear, starvation cactus
Synonym: *O. trichophora*
Look-alike Species: *O. polyacantha* var. *arenaria*

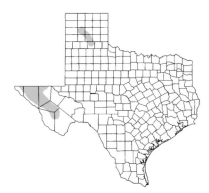

Opuntia pottsii — Potts' prickly pear

Features
Opuntia pottsii is recognized as a clump-forming upright shrub to 30 cm. It is one of the smallest prickly pears, frequently unnoticed in grasses until flowering. It is the only West Texas prickly pear with red flowers. Roots are often tuberlike. Pads commonly number only 6–10, are dark green, glaucous, obovate to diamond shaped, 5–20 cm long and 4–8 cm wide. Areoles are 3.0–3.5 mm, oval to subcircular, 4–6 per diagonal row at midstem, with tan wool. Spines are found on the upper half of pads.

Spines
Central: 1–4 erect, deflexed or projecting, twisted, 6 cm.
Color: Grayish white, black speckled, or reddish brown or purplish black with white tips.
Radial: Most always absent; if present, tiny and bristlelike.
Color: As the centrals.
Glochids: Dirty yellow to reddish brown, aging to dull brown in tuft, 3–5 mm.

Flowers
Totally red or orange flowers or yellow with red centers, 4–6 cm. Filaments: pale yellow, greenish yellow, or purplish; Anthers: yellow; Stigma: cream; Style: pinkish.

Fruits
Green to yellowish to dull red, elongate to obovoid, nearly spineless fruits, 2.5–4.0 cm.

Seeds
Tan to gray, subcircular, thick discoid seeds, 4–5 mm.

Habitat
This smallest prickly pear is found in sand, loam, gypsum, or limestone of grasslands, plains, and hills. Its range is from the southern Panhandle into the Trans-Pecos from the Pecos River west to El Paso County at elevations of 900–1,900 m (2,800–5,900 ft). Texas Areas 7–10.

Flowering Season
This species is expected to bloom from April through June. Flowers open about midday, close in midafternoon, and do not reopen. Flowering is complete in about one week.

Notes
Other Common Name: Potts' prickly pear
Synonyms: *O. macrorhiza* var. *pottsii*, *O. stenochila*
Look-alike Species: None

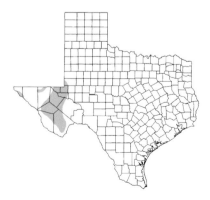

Opuntia pusilla   Sandbur prickly pear

Features
Opuntia pusilla is a very prostrate and spreading shrub with thin, easily detached, elongate stems in short trailing mats, seldom more than 10 cm tall but as wide as 4.5 m. Roots are entirely fibrous or with tubers. Stems are smooth, deep to light green with bluish color around the areoles, elliptic to linear, 2.5–5.0 cm long and 1.2–2.5 cm in diameter. Areoles are 3 mm, oval to subcircular in upper two-thirds of stem segment, with tan or gray wool. They wrinkle when dry but are not tuberculate. Stem branches are easily detached, becoming impaled in swimmer's feet on the beaches.

Spines
Central: 1–4 straight, flattened or acicular, deflexed or porrect, 3 cm.
Color: Light reddish brown, aging to gray with yellow tips.
Glochids: Greenish yellow to straw, in crescent of areole, 3 mm.

Flowers
Yellow to greenish yellow flowers, sometimes with deeper yellow centers, 2–3 cm. Filaments: yellow; Anthers: yellow; Stigma: white; Style: white.

Fruits
Red to purple, barrel-shaped fruits, 1.8–3.0 cm, with 8–16 spineless areoles.

Seeds
Tan, subcircular seeds, 4–6 mm.

Habitat
This prickly pear is found on sand dunes and rocky outcrops, behind the beaches along the Gulf Coast on Bolivar Peninsula at Galveston Bay in Galveston County at elevations of 0–50 m (0–165 ft). Texas Area 2.

Flowering Season
This species is expected to bloom from April through May.

Notes
Other Common Names: Cocklebur cactus, cock-spur cactus, crow-foot prickly pear, sandbur cactus
Synonyms: *Cactus pusillus*, *O. drummondii*
Look-alike Species: None.

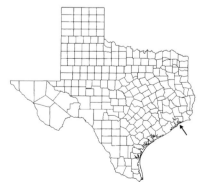

Opuntia rufida — Blind prickly pear

Features
Opuntia rufida is a conspicuously large erect shrub or tree. It is a spineless prickly pear with a short trunk, forming clumps common in desert flats, along south-facing cliffs, and on hillsides of dark, rough boulders. Many-branched, blue-green to grayish stems are velvety, circular, elliptic or obovate, 10–18 cm long and 5–25 cm wide. Branches are not easily disarticulating. Areoles are 3.0–3.2 mm, circular, 8–13 per diagonal row on midstem segment, with tiny white hairs and white or tan wool, aging to gray, 1.0–2.5 mm.

Spines
Central: None.
Radial: None.
Glochids: Reddish brown to white, numerous, crowded in areole, 1.0–2.5 mm.

Flowers
Yellow flowers, turning golden yellow to orange, 2.5–3.0 cm. Filaments: whitish; Anthers: yellow to cream; Stigma: dark green; Style: colorless.

Fruits
Red, subspherical to obovate, spineless fruits, 2.0–3.5 cm.

Seeds
Tan, irregularly discoid seeds, 2–3 mm.

Habitat
Blind prickly pear is found in sandy to gravelly desert soils of calcareous to volcanic material on desert hillsides along the Rio Grande flats, from southern Hudspeth to southern Brewster counties at elevations of 600–1,300 m (1,800–4,000 ft). Texas Areas 7 and 10.

Flowering Season
This cactus is expected to bloom in March and into May. Flowers open mid- to late morning, close at night, and often do not reopen.

Notes
Blind prickly pear derives its common name from the belief that the many loosely attached glochids will blow in the wind and injure the eyes of cattle.
Other Common Names: Blind pear, cinnamon pear, nopal cegador
Synonyms: *O. microdasys* var. *rufida*, *O. rufida* var. *tortiflora*
Look-alike Species: None

Opuntia spinosibacca — Spiny-fruited prickly pear

Features
Opuntia spinosibacca is an erect, branching, and sometimes spreading shrub growing to 1.5 m. A most stunning feature is its reddish spines and frequent purple blotches on the pads near the areoles. Glabrous pads are light yellowish, aging to green, and are ovate to circular, 10–25 cm long and 7.5–15.0 cm wide. Areoles are oval to circular, often with gray hair, and are found on low tubercles, a feature unusual among opuntias.

Spines
Central: 1–8 stout, flattened, twisted, and curved, 2–7 cm.
Color: Reddish to dark black with lighter tips.
Radial: 2–4 slender and acicular, 5–11 mm.
Color: Yellow to gray.
Glochids: Light brown with yellow tips in dense tufts.

Flowers
Golden yellow to orange flowers with red centers, 5–7 cm. Filaments: cream to pale green; Anthers: pale yellow; Stigma: pale green to pale yellow; Style: white to pink.

Fruits
Tan to reddish, round to ovoid and spiny fruits, 2.5–4.5 cm.

Seeds
Tan, disk-shaped seeds, 4–6 mm.

Habitat
This species grows on limestone hills and slopes in a small area of southeastern Big Bend National Park in Brewster County at elevations of 500–750 m (1,500–2,300 ft). Texas Area 10.

Flowering Season
Spiny-fruited prickly pear is expected to bloom during the day from April through May. Individual flowers may open and reopen for 2–3 days.

Notes
Other Common Name: Red-spined prickly pear
Synonym: *O. phaeacantha* var. *spinosibacca*
Look-alike Species: *O. camanchica*, *O. phaeacantha*

Opuntia stricta   Pest prickly pear

Features
Opuntia stricta is an erect, many-branched, and spreading shrub growing to 1 m. Glabrous, bluish green, aging to light green pads are elongated ovate to narrowly elliptic, 10–25 cm long and 7.5–15.0 cm wide. Areoles are elongated, raised ovals, 3.0–6.5 mm with 3–5 per diagonal row on midstem segment, with dense tan wool. Areoles on pad margins are somewhat slightly tuberculate, making pad edges appear scalloped.

Spines
Central: 0–3 acicular, straight or curved, spreading, 1.2–4.0 cm.
Color: Yellow aging to mottled brown; sometimes absent.
Glochids: Tan or yellow, aging to brown, inconspicuously few in adaxial crescent, to 4 mm.

Flowers
Light yellow flowers, 7.5–10.0 cm. Filaments: yellow; Anthers: yellow; Stigma: greenish white; Style: greenish yellow.

Fruits
Purplish, ellipsoid to barrel-shaped spineless fruits, 4–6 cm, with 6–10 areoles.

Seeds
Tan, subcircular seeds, 4–5 mm.

Habitat
This species is believed to be introduced in Texas on coastal sand dunes and shell middens only on Galveston Island in Galveston County near sea level. Texas Area 2.

Flowering Season
Pest prickly pear is expected to bloom from February to July.

Notes
Other Common Name: Cactus inermis
Synonyms: *Cactus strictus, O. stricta* var. *dillenii*
Look-alike Species: *O. engelmannii* var. *lindheimeri*

Opuntia strigil — Marble fruit prickly pear

Features
Opuntia strigil is an upright, compact, uncommonly sprawling shrub, to 1 m tall. Green, obovate to circular pads, 10–17 cm long and 8.5–15.0 cm wide. Spines produced over the entire pad except for the base. Areoles are 3–5 mm, oblong to elliptic, with 7–10 per diagonal row across midstem segment, with yellow-brown to brown wool. Fruits are the smallest of the prickly pears of the Southwest.

Spines
Central: 7–13, one acicular and porrect, others straight and deflexed, 1–4 cm.
Color: Reddish brown, reddish, or orange with yellow tips.
Radial: 2–4 acicular, deflexed, spreading, to 2 cm.
Color: Yellowish or reddish brown with yellow tips.
Glochids: Yellowish to reddish tan, erect and crowded in apical tuft, to 3 mm.

Flowers
Cream to lemon yellow flowers with orange midvein area, 2–3 cm. Filaments: cream yellow; Anthers: pale yellow; Stigma: pale green to cream; Style: pale cream.

Fruits
Red, subspherical, spineless fruits 1.5–2.8 cm, with 24–36 areoles.

Seeds
Tan, subcircular to reniform seeds, 3–4 mm.

Habitat
This prickly pear is found in shallow soils of limestone hills in desert scrub from the West Stockton Plateau in Reeves, Pecos, Terrell, and Val Verde counties, to Webb County near Laredo, at elevations of 100–200 m (330–660 ft). Texas Areas 6–8 and 10.

Flowering Season
This cactus is expected to bloom in April and May. Flowers open about midday to afternoon, close at night, and do not reopen.

Notes
Other Common Name: Bearded prickly pear
Synonyms: None
Look-alike Species: *O. atrispina* in the western part of its range

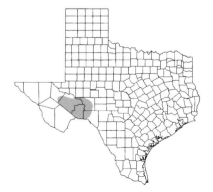

Opuntia tortispina — Twisted spine plains prickly pear

Features
Opuntia tortispina is a low, sprawling shrub to 40 cm tall, creeping and forming clumps two pads high. Thickened rootstocks produce branching stems with glossy, pale to deep green, broadly obovate to ovate pads, 6.5–15.0 cm long and 4–10 cm wide. Pads gray with age and are often wrinkled. Areoles are oval to subcircular, 2.5–5.0 mm, with 6–9 per diagonal row across midstem segments, with tan wool, aging to brown. Spines found on upper half of stem segment.

Spines
Central: 4–5, two terete or deflexed, and two to three acicular, projecting, 1.5–7.5 cm.
Color: Chalky white to gray with pale brown or yellowish tips.
Radial: 2 slender, acicular, spiral twisted, strongly deflexed, 5–15 mm.
Color: White with yellowish tips.
Glochids: Yellow to brownish white, well developed in tuft, to 6 mm.

Flowers
Yellow to gold flowers with distinct red to rusty red centers, 4–6 cm. Filaments: pale yellow with reddish tips; Anthers: Yellow; Stigma: greenish; Style: white to pale green.

Fruits
Reddish purple, oval to subspherical, spineless fruits, 3 cm, with 18–30 areoles.

Seeds
Whitish tan, irregular, flattened, 4–6 mm.

Habitat
This species is found in igneous or limestone alluvium, sand or sandy loam, in grasslands or pinyon-juniper woodlands. Its range extends from El Paso County to Brewster County and north through the Panhandle into New Mexico and Oklahoma at elevations of 1,400–1,800 m (4,300–5,600 ft). Texas Areas 8–10.

Flowering Season
This cactus is expected to bloom from May through June. Flowers open about midday, close at night, and do not reopen.

Notes
Other Common Name: Plains prickly pear
Synonyms: *O. cymochila, O. mackensenii, O. tenuispina, O. tortispina* var. *cymochila*
Look-alike Species: *O. macrorhiza, O. phaeacantha*

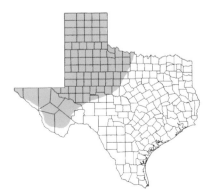

Acanthocereus tetragonus — Triangle cactus

Features
Acanthocereus tetragonus is an erect, coarsely shrubby, and sprawling plant found clambering among other vegetation and arching unless supported. Stems often branch near the base and sometimes root at the tips. The main stems grow from diffuse, fibrous roots. The dark green stems are noticeably angled, branching one or two times or more. Stem segments may be 20–300 cm long and 2.5–5.0 cm in diameter. Entire stems may reach 6 m in length. Stems have 3–5 sharply angled, winglike ribs with round, widely spaced areoles with short, whitish wool.

Spines
Central: 1–3 straight, porrect, and slightly deflexed, 1.8–4.0 cm.
Color: Light brown to gray or white.
Radial: 5–7 acicular or slightly flattened with bulbous bases, 6–25 mm.
Color: As the centrals.

Flowers
Large and showy nocturnal flowers are white and grow laterally on the stem near the terminal ends, 10–20 cm in diameter. Filaments: white; Anthers: light yellow; Stigma: white.

Fruits
Bright, shiny red, elliptic to ovate fruits bear spines on low tubercles, 3–8 cm. Split when ripe.

Seeds
Shiny black to brown, obovate, slightly keeled seeds, 4.4–4.8 mm.

Habitat
This species is found in sandy soils of dense bottomland thickets near coastal areas in Kenedy, Willacy, Hidalgo, Cameron, and Webb counties in the Rio Grande Valley at elevations of 0–10 m (0–33 ft). Texas Areas 2 and 6.

Flowering Season
This cereus blooms at night during midsummer through fall.

Notes
Other Common Names: Barbwire cactus, night-blooming cereus
Synonyms: *A. pentagonus, Cereus pentagonus*
Look-alike Species: *Echinocereus berlandieri, E. pentalophus*, though both species are much smaller

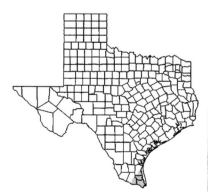

Echinocereus pentalophus — Lady finger cactus

Features
Echinocereus pentalophus is a branching and clustering cactus, forming low, disorganized clumps to 1 m across. Weakly erect or sprawling stems soon become decumbent. Light green to yellow-green stems, 10–60 cm long and 1–6 cm in diameter. The 4–5 ribs are sharp and straight to poorly defined and undulate. Low tubercles, more pronounced toward stem tips. Areoles have yellowish wool when young, naked thereafter.

Spines
Central: 0–1 terete, porrect to ascending, stiff, 1–3 cm.
Color: Yellowish tan, ashy white to dark gray, dark tipped.
Radial: 3–6 spreading, 2–20 mm.
Color: Reddish to tan with dark tips, aging to gray.

Flowers
Subapical, broadly funnelform, brilliant pink or magenta flowers with white or yellow throat, 1.5–7.5 cm. Filaments: greenish; Anthers: yellow-orange; Stigma: dark olive green.

Fruits
Green, ovoid fruits covered with brownish spines and white wool, 1.5–2.5 cm.

Seeds
Black, ovoid seeds, 1 mm.

Habitat
Lady finger cactus is found growing on limestone cliffs and alluvial coastal plains with lechuguilla in Tamaulipan thorn scrub in Cameron, Hidalgo, and Starr counties at elevations of 0–220 m (0–720 ft). Texas Areas 2 and 6.

Flowering Season
This species is expected to bloom from late March through early April. Flowers open about midday, close in late afternoon, and may reopen for 2–3 days.

Notes
This cactus is a most common species on both sides of the river in the Lower Rio Grande Valley.
Other Common Name: Alicoche
Synonym: *Cereus pentalophus*
Look-alike Species: *E. berlandieri*

Peniocereus greggii — Desert night-blooming cereus

Features
Peniocereus greggii is a long, slender, erect, or sprawling plant sheltered among the branches of a desert scrub. Turnip-shaped taproots are often quite large. Green to gray-green or purplish stems are slender and angular with 4–6 winged ribs, 15–30 cm long or more and 1–2 cm in diameter. Circular to elliptic areoles are on small tubercle-like projections on the rib crest, with white wool in new growth. Plants may branch.

Spines
Central: 1–2 porrect or deflexed, 1 mm.
Color: Black, aging to gray.
Radial: 6–9 appressed, slender, 3 mm.
Color: As the centrals.

Flowers
Laterally on stem from year-old areoles, flowers are white to red with greenish midribs, 5–8 cm. Filaments: cream; Anthers: pale yellowish tan or creamy yellow; Stigma: white.

Fruits
Bright red, ellipsoid fruits, 6–9 cm.

Seeds
Black, obovoid seeds, 3–4 mm.

Habitat
This cactus is found growing in nurse plants of acacia and other desert scrubs in desert flats, degraded grasslands, and alluvial basins. It is widespread but fairly uncommon in the Trans-Pecos. It grows at elevations of 1,000–1,500 m (3,300–5,000 ft). Texas Areas 7 and 10.

Flowering Season
This cereus is expected to bloom from April through May. Individual flowers open at dusk and close by early morning for only one night.

Notes
This cactus is highly camouflaged and combined with its scarcity, it becomes a most difficult specimen to locate.
Other Common Names: Chaparral cactus, queen of the night
Synonyms: *Cereus greggii, C. pottsii*
Look-alike Species: None

Echinocereus berlandieri — Berlandier's hedgehog cactus

Features
Echinocereus berlandieri is a sprawling and clustering plant with many branches. Stem tips are upright when young but lie prostrate when they increase in length. Diffuse roots. Deep to bright green, cylindrical stems, to 15 cm long and 1.5–2.5 cm in diameter. There are 4–6 ribs with undulate crests and conical tubercles with round, 3 mm areoles.

Spines
Central: 1–3 straight, stiff, ascending or projecting outward, 1.2–3.5 cm.
Color: White, brown bases, dark tipped.
Radial: 6–8 straight, appressed, spreading, 1 cm.
Color: Yellowish white, brown bases, dark tipped, aging to gray.

Flowers
Near apical, funnelform, 7.5–12.5 cm. Rose-pink to magenta with darker midstripes, throat bases white, petals in one row, long and narrow with pointed tips. Filaments: pale greenish; Anthers: yellow; Stigma: green; Style: green.

Fruits
Green to brownish, ovoid fruits, spines persistent, little wool, 2–3 cm.

Seeds
Black, ovoid seeds, 1.5 mm.

Habitat
This species inhabits shady thickets, from the coastal plains near the mouth of the Nueces River at Corpus Christi to the Lower Rio Grande Valley from elevations of 0–100 m (0–330 ft). Texas Areas 2 and 6.

Flowering Season
This plant is expected to bloom from May through June.

Notes
Other Common Names: None
Synonym: *Cereus berlandieri*
Look-alike Species: None

Echinocereus poselgeri — Dahlia cactus

Features
Echinocereus poselgeri is a very tall and slender plant that clambers through neighboring shrubs. Deep, tuberous roots initially give rise to erect stems that later become straggling or sprawling. Long, cylindrical dark blue-green stems turn brown and woody in the lower portion, 30–120 cm long and 0.6–1.5 cm in diameter. Plants sometimes have only a few branches with 8–10 low and inconspicuous ribs with interrupted crests and very small areoles, with some white wool.

Spines
Central: 1 straight, slightly flattened and upward pointing, 4–9 mm.
Color: Dark brown to black.
Radial: 8–16 straight, slender, and closely appressed, 1.5–4.5 mm.
Color: White to gray, tipped with brown or all dark.

Flowers
Near apical, funnelform rose-pink to magenta flowers with darker midstripes, 3.5–7.0 cm. Filaments: light yellow; Anthers: yellow; Stigma: green.

Fruits
Dark green to brownish, ovoid fruits, 2–3 cm. Wool and spines persistent.

Seeds
Black, ovoid seeds, 1.5 mm.

Habitat
This cactus inhabits alluvial, sandy soils in Tamaulipan thorn scrub in the Lower Rio Grande Valley from Cameron County northwest to Webb County near Laredo at elevations of 0–200 m (0–650 ft). Texas Areas 2 and 6.

Flowering Season
This species is expected to bloom from March through April, flowering near midday, closing before night, and reopening for 2–3 days.

Notes
Inconspicuous and camouflaged, this cactus is difficult to spot, as it often grows entwined among other shrubs for support.
Other Common Names: Pencil cactus, sacasil
Synonyms: *Cereus poselgeri, Wilcoxia poselgeri*
Look-alike Species: None

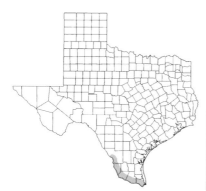

Coryphantha macromeris Big-needle pincushion cactus

Features
Coryphantha macromeris is a cespitose cactus, forming many-stemmed low mats or mounded clusters up to 100 cm wide. Massive, succulent roots produce profusely branching dark green to gray-green, deep-seated, hemispheric to short cylindrical stems 5–23 cm long and 4–8 cm in diameter. Distinctively large tubercles have areolar grooves distally more or less half their length, 1.5–3.8 cm. Circular, 4 mm areoles are found at the tip of each tubercle, with white wool disappearing with age.

Spines
Central: 3–8 slightly curved, 3–7 cm.
Color: Dark gray to black.
Radial: 9–15 flattened, 1.5–2.5 cm.
Color: Tan to brown, aging gray to white.

Flowers
Nearly apical rose, pink, or magenta flowers with a darker midstripe, 3–7 cm.
Filaments: greenish white; Anthers: bright yellow; Stigma: white or pale yellow.

Fruits
Dark green ovoid or ellipsoid fruits, 1.4–3.0 cm.

Seeds
Reddish brown, comma-shaped or spherical seeds, 1.2–1.5 mm.

Habitat
This cactus is one of the most widespread cacti in the Trans-Pecos. It is found in sandy alluvium, gravelly benches, or clay in Chihuahuan desert scrub and Tamaulipan thorn scrub. It is also found in Starr County in the Rio Grande Valley at elevations of 30–2,000 m (100–6,300 ft). Texas Areas 6, 7, 9, and 10.

Flowering Season
These plants can be expected to bloom from June to August in the afternoon for only one day.

Notes
This cactus is found most frequently under the shade of shrubs and trees, where it may form large mats of many plants.
Other Common Names: Big nipple cory cactus, long mamma, nipple beehive cactus
Synonyms: *Echinocactus macromeris*, *Mammillaria macromeris*
Look-alike Species: None

Coryphantha macromeris var. *runyonii*   Runyon's coryphantha

Features
Coryphantha macromeris var. *runyonii* is a distinct South Texas form of *C. macromeris* of the Trans-Pecos. It, too, is a cespitose cactus, forming irregular mounds or mounded clusters. Succulent roots produce profusely branching dark green to gray-green, deep-seated, hemispheric to short cylindrical stems 5–23 cm long and 4–8 cm in diameter. Distinctively large tubercles have areolar grooves distally more or less half their length, 1–2 cm. Circular areoles with white wool are found at the tip of each tubercle.

Spines
Central: 3–8 slightly curved, 3–7 cm.
Color: Dark gray to black.
Radial: 9–15 flattened, 1.5–2.5 cm.
Color: Tan to brown, aging gray to white.

Flowers
Nearly apical rose, pink, or magenta flowers with a darker midstripe, 3–7 cm. Filaments: greenish white; Anthers: bright yellow; Stigma: white or pale yellow.

Fruits
Dark green ovoid or ellipsoid fruits, 1.4–3.0 cm.

Seeds
Reddish brown, comma-shaped or spherical seeds, 1.2–1.5 mm.

Habitat
This cactus is found within a small area along the Rio Grande in the Lower Rio Grande Valley. It grows on gravelly hillsides and in silt in Tamaulipan thorn scrub. It is found in Starr County in the Rio Grande Valley at elevations of 30–100 m (100–300 ft). Texas Areas 2 and 6.

Flowering Season
These plants can be expected to bloom from June through August in the afternoon for only one day.

Notes
This cactus is found most frequently under the shade of shrubs and trees, where it may form large mats of many plants.
Other Common Names: Big nipple cory cactus, long mamma, nipple beehive cactus
Synonyms: *C. macromeris*, *Mammillaria macromeris*
Look-alike Species: None

Coryphantha robustispina var. *scheeri* — Pineapple cactus

Features
Coryphantha robustispina var. *scheeri* has hooked central spines (eastern range, straight) that do not obscure the stem surface. Diffuse roots develop deep-seated stems when young. Stems are dull gray-green, globose to ovoid to cylindrical; sometimes flat-topped, 5–15 cm long and 5.5–8.5 cm in diameter. Stems are branching, rarely forming mounds, 40 cm across. Tubercles are conical and firm with grooves extending the full length of the tubercle, 1.5–3.0 cm. Areoles are oval and white with areolar glands present.

Spines
Central: 1–5 straight or hooked downward, stout with large bulbous base, 2.3–3.4 cm.
Color: Gray with red-brown tips.
Radial: 6–16 terete or dorsoventrally flattened, appressed and spokelike, stout with large bulbous base, 1.1–3.5 cm.
Color: Pale gray to white with dark tips.

Flowers
Nearly apical, dark golden brown to greenish yellow flowers are sometimes reddish at the center with vague reddish midstripes, 4.5–6.4 cm. Filaments: reddish orange; Anthers: yellow; Stigma: cream to orange; Style: orange to cream.

Fruits
Green, cylindrical fruits, 4–5 cm.

Seeds
Bright reddish brown, reniform or comma-shaped seeds, 2.3–3.5 mm.

Habitat
This cactus is found in deep, sandy soils of sedimentary or igneous rocks in oak-juniper savannahs, grassy hills, and valley floors in creosotebush desert scrub from El Paso to Hudspeth counties at elevations of 900–1,800 m (3,000–6,000 ft). Texas Area 10.

Flowering Season
This species is expected to bloom opportunistically multiple times from April through September, with the blooms lasting one day.

Notes
Other Common Name: Needle "mulee" beehive
Synonyms: *C. scheeri*, *Mammillaria robustispina*
Look-alike Species: Young adult plants of *Ancistrocactus brevihamatus*

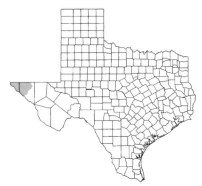

Echinocereus chisosensis — Chisos Mountain hedgehog cactus

Features
Echinocereus chisosensis is an inconspicuous, solitary cactus or branched, forming clumps. Tuberous roots produce light green to gray-green, erect, cylindrical stems that taper toward the apex, 12.5–20.0 cm long and 3–5 cm in diameter. Stems sometimes branch in older plants, with 10–16 ribs with strongly tuberculate crests. Tubercles slender, areoles on upper stem prominent with puffy white wool, others circular and mostly naked.

Spines
Central: 2–4 terete, appressed to slightly spreading, 3.5–17.0 mm.
Color: Dark gray to brown or purplish black, sometimes with annual rings.
Radial: 10–17 straight, appressed, divergent, 6–10 mm.
Color: Tan or ashy white to pinkish gray, some with red-brown tips.

Flowers
Magenta funnelform flowers with crimson centers and rose-pink tips, 5–7 cm. Filaments: white to pink; Anthers: pale yellow; Stigma: green; Style: white.

Fruits
Green to dull red, oblong to obovoid fruits covered with wool and spines, 1.5–3.5 cm.

Seeds
Black, ovoid and strongly tuberculate seeds, 1.2 mm.

Habitat
This species is found in gravel and sandy alluvium in Chihuahuan desert scrub, in low vegetation in grasses, in lechuguilla, or in mats of dog cholla. Unique to southern Big Bend National Park in Brewster County at elevations of 600–900 m (2,000–3,000 ft). Texas Area 10.

Flowering Season
This hedgehog cactus is expected to bloom in mid-March through mid-April in midmorning and afternoon, sometimes for 1–3 days.

Notes
This uncommon cactus is difficult to locate because it is always found sheltered in the protection of nurse plants.
Other Common Names: Chisos hedgehog cactus, Chisos pitaya
Synonym: *E. reichenbachii* var. *chisosensis*
Look-alike Species: *E. dasyacanthus, E. viridiflorus*

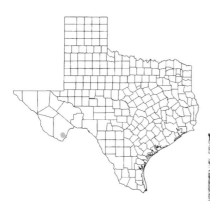

Echinocereus coccineus — Claret cup cactus

Features
Echinocereus coccineus is a commonly cespitose, clustering cactus that is widely variable in form throughout its range. The species frequently forms 20–100 branched clumps or rounded mounds up to 1 m across. Erect, green, cylindrical or spherical stems are 5–40 cm long and 4–14 cm in diameter with 7–9 ribs with undulate crests. Round areoles bear spines that partially obscure stems.

Spines
Central: 0–6, commonly one terete with bulbous bases, 1–8 cm.
Color: Ashy white to gray with dark tips.
Radial: 4–13 mostly straight, 5–40 mm.
Color: As the centrals.

Flowers
Waxy, crimson or scarlet funnelform flowers, 3–7 cm. Filaments: cream to pinkish; Anthers: pink or purple; Stigma: green.

Fruits
Greenish fruits, turning yellowish pink, ripening to pale orange to brick red, bearing spine clusters, 2–4 cm.

Seeds
Black, globose papillate seeds, 1.3–2.0 mm.

Habitat
This claret cup is found in igneous, metamorphic, and limestone substrates in Chihuahuan desert scrub, grasslands, and pinyon and oak-juniper woodlands throughout the Trans-Pecos to Central Texas and the Hill Country at elevations of 150 to 2,700 m (500–8,800 ft). Texas Areas 6–8 and 10.

Flowering Season
This species is expected to bloom in March and April but may be delayed until May in times of freezing weather. Flowers remain open at night for 3–4 days or longer.

Notes
Other Common Names: Langtry claret cup cactus, pitahaya, scarlet hedgehog cactus, strawberry cactus, Texas claret cup cactus
Synonyms: *E. coccineus* var. *aggregatus*, *E. triglochidiatus* var. *paucispinus*
Look-alike Species: *E. fendleri*

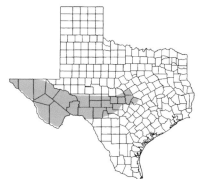

Echinocereus dasyacanthus — Texas rainbow cactus

Features
Echinocereus dasyacanthus has spines that present a bristly appearance with subtle annual rings of contrasting "rainbow" bands of spine colors. Roots are diffuse to tuberous. Erect, mostly solitary stems, or with 2–3 basal branches, a few forming clumps of less than 20 branches in old age. Stems are ovoid to cylindrical, 11–23 cm long and 5.5–7.0 cm in diameter. The 15–19 ribs have sharply undulate crests.

Spines
Central: 4–12 terete, acicular, spreading, darker than radials, 5–12 mm.
Color: Tan or pale yellow to pink, infrequently ashy white to reddish brown, tips often dark.
Radial: 14–25 acicular, straight, appressed, overlapping those of adjacent rows, 7–20 mm.
Color: Tan or pale yellow to pink, infrequently ashy white to reddish brown, tips often dark.

Flowers
Showy, polymorphic, near apical, yellow to magenta or cyanic orange flowers with green centers, 7–12 cm. Filaments: pale green; Anthers: yellow; Stigma: green; Style: white.

Fruits
Green, turning dark purplish to maroon, globose to ellipsoid fruits. Bear spiny, deciduous areoles, 3–6 cm.

Seeds
Black, globular, papillate seeds, 1.0–1.4 mm.

Habitat
Rainbow cactus habitat includes limestone soils in valleys with Chihuahuan desert scrub to rocky canyon sides, throughout the Trans-Pecos region except Val Verde County, at elevations of 600–1,500 m (2,000–5,000 ft). Texas Areas 6–10.

Flowering Season
This cactus is expected to bloom from March through May. Flowers open in the morning and afternoon, close at night, and reopen for 3–7 days.

Notes
Other Common Names: Golden rainbow hedgehog cactus, rainbow cactus, spiny hedgehog cactus
Synonym: *E. pectinatus* var. *dasyacanthus*
Look-alike Species: *E. viridiflorus*

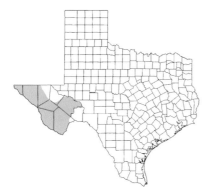

Echinocereus enneacanthus var. *brevispinus* Strawberry pitaya

Features
Echinocereus enneacanthus var. *brevispinus* is a branching cactus forming flat-topped or crowded clumps with 15–100 stems or more. Green, slender, and upright; sometimes prostrate or decumbent stems are 8–40 cm long and 3–5 cm in diameter. The 7–9 ribs have uninterrupted crests. Circular areoles have short matted or woolly hairs.

Spines
Central: 1–4 porrect, straight, angular or flattened, slender, 2.0–4.4 cm.
Color: Yellowish, brownish, or blue-gray.
Radial: 8–10 straight with bulbous bases, 9.5–19.0 mm.
Color: Whitish to tan or brownish, with darker tips.

Flowers
Subapical, pink or magenta funnelform flowers with dark reddish throat, 7–10 cm. Filaments: greenish to pink; Anthers: yellow; Stigma: green.

Fruits
Greenish to dull red, ripening to red, globular to ovoid fruits, strawberry aroma, 2–3 cm.

Seeds
Black, ovoid, strongly tuberculate seeds, 1.0–1.4 mm.

Habitat
This species inhabits well-drained alluvium of limestone soils along limestone bluffs within Tamaulipan thorn scrub and oak savannahs in the eastern Trans-Pecos region to the Edwards Plateau, south to the Rio Grande Valley at elevations of 0–1,000 m (0–3,100 ft). Texas Areas 2, 6, 7, and 10.

Flowering Season
This cactus is expected to bloom from April through May. Flowers open in succession, partially close at night, and reopen for 2–4 days.

Notes
Other Common Names: Mexican strawberry cactus, mound pitaya, pitaya
Synonym: *E. enneacanthus* forma *brevispinus*
Look-alike Species: *E. enneacanthus* var. *enneacanthus*

Echinocereus enneacanthus var. *enneacanthus* — Strawberry hedgehog cactus

Features
Echinocereus enneacanthus var. *enneacanthus* is a somewhat flaccid, often sprawling and branching species, forming dense or loose clumps with 20–100 branches, usually branching before blooming. Medium to light green, cylindrical stems, the longest sometimes prostrate, 15 cm long and 5–15 cm in diameter. The 7–10 ribs have uninterrupted crests and circular areoles.

Spines
Central: 1–4 porrect, flattened, stout, and slightly curved, 5.5–9.0 cm.
Color: Opaque, white, tan, or gray, often nearly black.
Radial: 5–8 straight, flattened, 9–40 mm.
Color: Often tipped or banded with brown.

Flowers
Pink or magenta funnelform flowers with deep reddish throat, 7–11 cm. Filaments: greenish to pink; Anthers: yellow; Stigma: green; Style: whitish.

Fruits
Yellow-green or dull reddish, maturing to bright red, globular to ovoid fruits, 2–3 cm.

Seeds
Black irregular, globular, or ovoid tuberculate seeds, 1.0–1.4 mm.

Habitat
This hedgehog cactus is found on gravelly hillsides, alluvial washes, and flats among Chihuahuan desert scrub from El Paso and Hudspeth counties southeast to Brewster County at elevations of 600–1,800 m (2,000–6,000 ft). Texas Area 10.

Flowering Season
Strawberry hedgehog cactus is expected to bloom April through June during midmorning. Flowers close at night and reopen for 2–4 days.

Notes
Other Common Names: Alicoche, pitaya, strawberry hedgehog cactus
Synonym: *E. dubius*
Look-alike Species: *E. enneacanthus* var. *brevispinus*

Echinocereus fendleri — Fendler's hedgehog cactus

Features
Echinocereus fendleri is a branching species, sometime forming loose clumps of up to 20 branches. Roots are diffuse to tuberous. Dark green, erect or slightly decumbent and flaccid, ovoid to cylindrical, somewhat wrinkled stems 7.5–17.0 cm long and 3.8–7.5 cm in diameter. The 8–11 ribs have uninterrupted or undulate crests and circular areoles surrounded by conspicuous swellings. Spines only somewhat hide the stem.

Spines
Central: 1 straight or upward curving, porrect and bulbous at the base, 2.5–5.0 cm.
Color: Most commonly black.
Radials: 4–10 spreading, bulbous bases, 1.1–3.9 cm.
Color: Opaque white to ashy gray, often with a dark stripe.

Flowers
Large and showy, magenta or rarely pink or white flowers with darker purple throats, 5–11 cm. Filaments: greenish; Anthers: yellow; Stigma: dark green; Style: white.

Fruits
Bright red to brick red, ellipsoidal or almost spherical fruits, 2–3 cm. Areoles with white spines and wool are absent at maturity.

Seeds
Black, globular to obovoid reticulate seeds, 1.0–1.5 mm.

Habitat
This species is found in limestone or igneous substrates in pinyon-juniper woodlands, mesquite thickets, and semidesert grasslands in the western Trans-Pecos region from El Paso to Presidio counties at elevations of 900–1,500 m (3,000–5,000 ft). Texas Area 10.

Flowering Season
This cactus is expected to bloom from mid-April to mid-May in the morning and afternoon, partially closes at night, and reopens for 1–7 days.

Notes
Other Common Names: Fendler's pitaya, purple hedgehog cactus, strawberry cactus
Synonyms: *Cereus fendleri*, *E. fendleri* var. *rectispinus*
Look-alike Species: *E. coccineus*, *E. viridiflorus*

Echinocereus pectinatus var. *wenigeri*   Langtry rainbow cactus

Features
Echinocereus pectinatus var. *wenigeri* is entirely whitish in appearance without obvious color banding, with stems obscured by spines. Solitary stems or with 1–2 branches at the base in older plants. Stems erect, short, cylindrical or spherical, 8–17 cm long and 5.6–8.0 cm in diameter. The 13–18 ribs have slightly undulate crests and low tubercles with broadly oval areoles, with woolly hairs near the apex or elongated and naked lower on the stem.

Spines
Central: 1–6 smooth, terete, often in one vertical row, inconspicuous, less than 3 mm.
Color: Ashy white, tipped pink or purplish brown.
Radial: 14–20 pectinate, tightly appressed, recurved between ribs, 5–15 mm.
Color: Ashy white, tipped pink or purplish brown.

Flowers
Multicolored flowers from pink, golden yellow, or magenta with white or multicolor banded centers, 5.5–10.0 cm. Filaments: white; Anthers: yellow; Stigma: dark green.

Fruits
Green to purplish, turning reddish, globose to broadly elliptic fruits, 2.5–3.0 cm.

Seeds
Black, globular, papillate seeds, 1.0–1.3 mm.

Habitat
This species is found in limestone soils and outcrops in Chihuahuan desert scrub or poor grasslands and Tamaulipan thorn scrub in Terrell and Val Verde counties at elevations of 300–1,700 m (1,000–5,600 ft). Texas Areas 7 and 10.

Flowering Season
This plant is expected to bloom from March through May in the morning, closing at night and reopening again for 2–3 days.

Notes
Other Common Names: Ashy white pitaya, comb hedgehog, Weniger hedgehog
Synonym: *Echinocactus pectinatus*
Look-alike Species: *Echinocereus reichenbachii*

Echinocereus reichenbachii Lace cactus

Features
Echinocereus reichenbachii is easily identified by pectinately arranged, appressed radial spines in a single row. Diffuse roots support simple, sometimes branching stems with as many as 12 branches. Erect, light or dark green, cylindrical, sometimes tapered and constricted stems are partially obscured by spines 7.5–40.0 cm long and 4–10 cm in diameter. The 10–19 narrow and slightly undulate ribs have low tubercles and elliptic to linear areoles with white wool near stem apex.

Spines
Central: 0–7 slender, terete but usually lacking, 1–6 mm.
Color: Ashy white with dark tips.
Radial: 12–36 straight to slightly curved, appressed, and interlocking those of adjacent areoles, 2–8 mm.
Color: Ashy white with dark tips.

Flowers
Near apical, funnelform, pink to magenta flowers, frequently white, 5–10 mm. Flowers emerge from diffuse woolly areoles. Filaments: reddish; Anthers: creamy yellow; Stigma: green; Style: reddish to pinkish.

Fruits
Green, globose to ovoid fruits, covered with wool and slender spines, 1.5–2.5 cm.

Seeds
Ovoid, strongly tuberculate seeds, 1.5 mm.

Habitat
Lace cactus is found in rocky, sandy soils and limestone among desert scrubs, grasslands, and oak-juniper woodlands from the Hill Country and Central Texas south to the Rio Grande Valley, northward to the Panhandle, and west to the eastern edge of the Trans-Pecos at elevations of 0–1,800 m (0–6,000 ft). Texas Areas 3–10.

Flowering Season
This species is expected to bloom from April through June in the afternoon for one day only.

Notes
The lace cactus is perhaps one of the best-known cacti in our area. It is a common cactus of Oklahoma and Central Texas and is prolific in its range.
Other Common Names: Lace hedgehog cactus, purple candle
Synonym: *Echinocactus reichenbachii*
Look-alike Species: *Echinocereus pectinatus* var. *wenigeri*

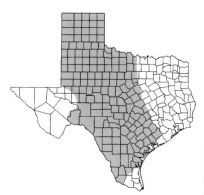

Echinocereus stramineus var. stramineus — Strawberry cactus

Features
Echinocereus stramineus var. *stramineus* is recognized as a cespitose, hemispheric mound bristling with unusually long, yellow-tipped spines that form clumps of as many as 20–100 or more branches 15–100 cm across. Erect, ovoid to cylindrical stems taper gradually to the apex, up to 30 cm long and 4.5–11.0 cm in diameter. The 10–17 ribs have slightly undulate crests with closely arranged, circular areoles producing dense spine growth that mostly obscures the stems.

Spines
Central: 2–4 terete or somewhat flattened, straight or slightly curved and strongly projecting, 5.7–9.0 cm.
Color: Translucent white to straw colored.
Radial: 7–14 acicular, somewhat flattened and straight, 1.5–4.0 cm.
Color: Translucent white to straw colored.

Flowers
Near apical, rose-pink to magenta flowers, some with darker throats, 8.5–12.0 cm. Filaments: reddish; Anthers: yellow; Stigma: green; Style: reddish.

Fruits
Reticulated, pinkish brown, globular to broad oval fruits, 3.5–4.0 cm. These have the color, smell, and flavor of strawberries, hence the common name.

Seeds
Black, ovoid, tuberculate seeds, 1.2–1.5 mm.

Habitat
This species is found on exposed limestone outcrops, igneous and sedimentary substrates, and rocky slopes among Chihuahuan desert scrub of the western and southern Trans-Pecos region from El Paso to the Pecos River at elevations of 500–1,800 m (1,650–6,000 ft). Texas Areas 7 and 10.

Flowering Season
This cactus is expected to bloom from late March through May, commonly with 20–30 stems flowering in one mound. Flowers close at night and reopen for 1–4 days.

Notes
From a distance the large mounds of this species appear quite like glistening piles of straw due to the messy appearance of the long, yellow spines. The sight of 20 flowers in full bloom on one mound is spectacular.
Other Common Name: Strawberry hedgehog
Synonyms: *Cereus stramineus*, *E. enneacanthus* var. *stramineus*
Look-alike Species: *E. enneacanthus* var. *enneacanthus*

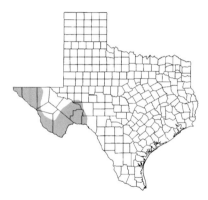

Echinocereus viridiflorus — Green-flowered hedgehog cactus

Features
Echinocereus viridiflorus is a highly variable complex of about 12 known varieties, many of which occur in Texas. The species has a red or reddish brown, variable to greenish yellow, tan, or white overall appearance. Erect, short cylindrical to ovoid stems, 8–30 cm long and 5–8 cm in diameter. The plants sometimes form up to 10–12 branches in old age. The 12–17 ribs with prominent and slightly undulate crests produce elliptic to oblong, 3–4 mm areoles with wool at the stem apex.

Spines
Central: 0–2 terete, straight and stiff, often one is conspicuously prominent, 6–20 mm.
Color: Gray or yellow basal halves with reddish distal halves, or whitish to yellow overall.
Radial: 14–23 pectinate, tightly appressed to spreading, with bulbous bases, 8–11 mm.
Color: Tan or yellowish, reddish, or proximally yellow with red distal portion, or ashy white with red tips.

Flowers
Produced about midstem, unscented funnelform flowers are amber, sulfur yellow, purplish red to brownish green, and may not open fully, 1.5–3.0 cm. Filaments: yellow; Anthers: yellow; Stigma: green; Style: green.

Fruits
Dark green to dull red, oval to circular, lemon-scented fruits, 9–17 mm.

Seeds
Black, circular seeds, 1.0–1.2 mm.

Habitat
This species is a familiar Trans-Pecos cactus. It is found in igneous rocky, gravelly, or silty alluvial substrates or novaculite in Chihuahuan desert scrub, grasslands, or shortgrass prairies in the Trans-Pecos, north through the Panhandle, and into adjacent states at elevations of 700–1,800 m (2,200–5,600 ft). Texas Areas 7–10.

Flowering Season
This cactus is expected to bloom from April through May and opportunistically thereafter. Plants may produce as many as 50 flowers in a season that last 12–20 days.

Notes
This is a small and often hard-to-see cactus. It is amazing to know how many plants may be hidden in the grass.
Other Common Name: Nylon hedgehog cactus
Synonym: *E. chloranthus* var. *cylindricus*
Look-alike Species: *E. dasyacanthus*

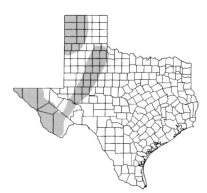

Echinocereus viridiflorus var. *canus*

Solitario green-flowered hedgehog cactus

Features
Echinocereus viridiflorus var. *canus* is usually a solitary plant identified by its shaggy appearance of white hairs in horizontal bands around the stem. Erect, ovoid to cylindrical stems, 6–15 cm long and 3–6 cm in diameter. Spines partially obscure stems with 14–16 ribs.

Spines
Central: 8–15 fine, flexible, and spreading or curving randomly from bulbous bases, 1.7–2.5 cm.
Color: White with reddish tips.
Radial: 30–50 appressed, 5–8 mm.
Color: White.

Flowers
Scented, light to golden green flowers are lateral on the stems, 2.0–3.5 cm. Filaments: greenish white; Anthers: yellow; Stigma: dark green.

Fruits
Green ovoid to oblong fruits obscured by 18 or more white spines in each areole.

Seeds
Black, pyriform seeds, 1.0–1.3 mm.

Habitat
This cactus inhabits caballos novaculite outcrops inside the Solitario Dome in southern Presidio County. The species is unique to this location at elevations of 1,300–1,500 m (4,400–4,800 ft). Texas Area 10.

Flowering Season
These plants are expected to bloom during March into May for 3–4 days.

Notes
Other Common Name: Davis hedgehog cactus
Synonym: *E. viridiflorus* var. *nova*
Look-alike Species: *Coryphantha scheeri* var. *albicolumnaria*, *Mammillaria pottsii*

Echinocereus viridiflorus var. *chloranthus* Western green-flowered hedgehog cactus

Features
Echinocereus viridiflorus var. *chloranthus* is usually a solitary or branching, erect and cylindrical cactus identifiable by its bristly appearance, horizontal color banding, and curved lower central spine. Older plants may exhibit limited branching with as many as 10 or more branches. It has stems, 7–20 cm long and 5–8 cm in diameter. Spines partially obscure stems with 11–16 well-defined undulate ribs.

Spines
Central: 2–6 straight with one often curved, 6–20 mm.
Color: Reddish or white with reddish tips.
Radial: 15–23 tightly appressed and spreading, 1.2–1.4 cm.
Color: Ashy white, reddish, or reddish black in horizontal bands around the stem.

Flowers
Unscented, dark green to greenish brown funnelform flowers are produced on lateral midstems, 2.0–3.4 cm. Filaments: pale yellow; Anthers: yellow; Stigma: green.

Fruits
Green, round fruits, 9–17 mm.

Seeds
Black, round, papillate seeds, 1.0–1.5 mm.

Habitat
This cactus inhabits arid areas in igneous and sedimentary soils from El Paso County east to Hudspeth and Culberson counties at elevations of 1,200–1,600 m (3,900–5,300 ft). Texas Area 10.

Flowering Season
These plants are expected to bloom March through May for 15–30 days.

Notes
Other Common Names: Cinder bells, green-flowered pitaya, green-flowered torch cactus
Synonyms: *E. chloranthus*, *E. chloranthus* var. *chloranthus*
Look-alike Species: *E. dasyacanthus*

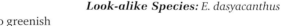

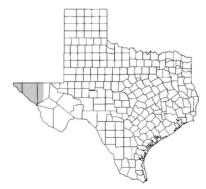

Echinocereus viridiflorus var. *correllii* Correll's green-flowered hedgehog cactus

Features
Echinocereus viridiflorus var. *correllii* is usually an unbranched plant with greenish yellow ovoid to short cylindrical stems, up to 20 cm long and 8 cm in diameter. Spines partially obscure stems with 14–19 ribs.

Spines
Central: 1–3 terete, one much longer than the others when present, 7–13 mm.
Color: Yellowish or white with reddish tips.
Radial: 8–12 slightly flattened and pectinately appressed, 8–12 mm.
Color: Gray or ashy white, brown or reddish tipped.

Flowers
Greenish yellow flowers arise mid-stem, 2.5–3.0 cm. Filaments: pale yellow; Anthers: light yellow; Stigma: green; Style: green.

Fruits
Green to dark red, ovate fruits, 5.5–9.0 cm.

Seeds
Black, nearly round seeds, 1 mm.

Habitat
This cactus inhabits novaculite soils on slopes and rocky hills in alluvial desert grasslands. The species is found in a few isolated sites in Brewster and Pecos counties at elevations of 100–1,400 m (300–4,000 ft). Texas Area 10.

Flowering Season
These plants are expected to bloom in April and May and may remain open overnight.

Notes
Other Common Name: Correll hedgehog cactus
Synonym: *E. viridiflorus* subsp. *correllii*
Look-alike Species: *E. viridiflorus* var. *cylindricus*

Echinocereus viridiflorus var. *cylindricus* — Small-flowered hedgehog cactus

Features
Echinocereus viridiflorus var. *cylindricus* is usually a solitary, erect, and cylindrical cactus identifiable by its red or red-brown and white appearance. Older plants may exhibit limited branching with as many as 10 or more branches. The stems are 8–30 cm long and 5–8 cm in diameter. Spines partially obscure stems with 12–17 ribs and slightly undulating, yet prominent crests.

Spines
Central: 0–2 straight with one often prominent, 6–20 mm.
Color: White with reddish tips.
Radial: 14–23 tightly appressed and spreading, 8–11 mm.
Color: White.

Flowers
Unscented, greenish brown, carmine, amber, or yellow funnelform flowers are produced on lateral midstems, 2.0–3.5 cm. Filaments: pale yellow; Anthers: yellow; Stigma: dark green.

Fruits
Dark green, oval to round fruits, 9–17 mm.

Seeds
Black, broad seeds, 1.0–1.2 mm.

Habitat
This cactus inhabits rocky and alluvial substrates in the western Trans-Pecos from El Paso County south to Brewster County at elevations of 700–2,700 m (2,300–8,900 ft). Texas Area 10.

Flowering Season
These plants are expected to bloom in April and May for 12–20 days.

Notes
Other Common Names: Green-flowered pitaya, New Mexico rainbow cactus, nylon hedgehog cactus
Synonyms: *E. chloranthus* var. *cylindricus*, *E. stanleyi*
Look-alike Species: *E. dasyacanthus*

Echinocereus x roetteri var. *neomexicana* Lloyd's hedgehog cactus

Features
Echinocereus x roetteri var. *neomexicana* is frequently found as an unbranched plant but with the ability to form clumps of many stems. Diffuse roots produce green, ovoid to cylindrical stems, 19–30 cm long and 6–10 cm in diameter. The 12 ribs are divided into distinct tubercles with oval areoles.

Spines
Central: 4–6 terete or flattened with bulbous base, 1.2–2.0 cm.
Color: Ash gray to dark brown, some with brownish tips.
Radial: 12–16 terete, 1.2–2.0 cm.
Color: As the centrals.

Flowers
Orange, to pink to yellow, 4.5–7.0 cm. Filaments: red; Anthers: red; Stigma: yellowish green; Style: pale green.

Fruits
Maroon to red fruits with circular areoles and white spines, 5.5–9.0 cm.

Seeds
Black, papillate seeds, 1.0–1.5 mm.

Habitat
This cactus inhabits hillsides or alluvial flats in mesquite and creosotebush desert scrub. The species is found in the central and southern Trans-Pecos at elevations of 750–1,360 m (2,500–4,500 ft). Texas Area 10.

Flowering Season
These plants are expected to bloom in April and May in the morning and afternoon for 3–5 days.

Notes
Other Common Name: Davis hedgehog cactus
Synonyms: *E. lloydii, E. x roetteri*
Look-alike Species: *E. coccineus, E. enneacanthus, E. fendleri*

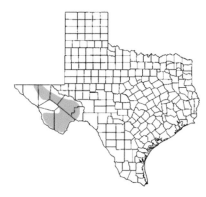

Escobaria dasyacantha var. *chaffeyi* — Chaffey's pincushion cactus

Features
Escobaria dasyacantha var. *chaffeyi* is a shaggy, mountain-dwelling cactus with spines that obscure the stems. Diffuse roots give rise to spherical to short cylindrical green stems that are 5.0–10.5 cm long and 3.0–4.5 cm in diameter. Stems branch infrequently, though 2–3 branches occur commonly in the Chisos Mountains. Stems produce 7–8 mm tubercles with circular areoles, 2.0–2.5 mm.

Spines
Central: 8–10 straight and spreading with bulbous bases, 7–15 mm.
Color: White with yellow or red at the tips.
Radial: 22–40+ straight, appressed, 5–9 mm.
Color: White to gray with yellow tips.

Flowers
Nearly apical, pink to creamy white flowers with prominent light brown midstripes, 6.5–13.0 mm. Filaments: pink; Anthers: yellow; Stigma: green; Style: greenish.

Fruits
Bright red to deep pink, obovoid to clavate fruits, 1.3–2.4 cm.

Seeds
Black, comma-shaped seeds, 1.0–1.2 mm.

Habitat
Chaffey's pincushion cactus is unique to limestone and igneous substrates within oak, juniper, and pine woodlands, among rocks in Selaginella spikemoss mats and grassy areas of the Upper Chisos Mountains of Big Bend National Park in Brewster County at elevations of 1,500–2,000 m (4,600–6,200 ft). Texas Area 10.

Flowering Season
These plants bloom from March through June in the mornings for 2–4 days.

Notes
Other Common Name: Biscuit cactus
Synonyms: *Coryphantha chaffeyi*, *E. chaffeyi*
Look-alike Species: *E. dasyacantha*, *E. duncanii*

Escobaria dasyacantha var. *dasyacantha*   Desert pincushion cactus

Features
Escobaria dasyacantha var. *dasyacantha* is a small, cryptic cactus with a shaggy overall appearance, its diffuse spines obscuring the stems. Diffuse roots give rise to globular to cylindroid green stems 4.5–10.0 cm long and 3.0–4.5 cm in diameter, but up to 15–17 cm long in the shade. Solitary, rarely branched up to 2–5 stems in the shade. Tubercles are 7–8 mm with a felt-filled groove supporting circular, 2.0–2.5 mm areoles with white wool in young areoles.

Spines
Central: 4–9 porrect with bulbous bases angled diffusely from stem, 1.2–1.7 cm.
Color: White with reddish brown to black tips.
Radial: 21–31 appressed, 6–9 mm.
Color: As the centrals.

Flowers
Nearly apical flowers are the smallest of the genus. Light pink to creamy white, 12 mm. Filaments: white to colorless; Anthers: yellow; Stigma: green; Style: greenish.

Fruits
Bright red clavate to cylindrical fruits, 1.3–2.7 cm.

Seeds
Black pitted, subspherical or comma-shaped seeds, 1.0–1.2 mm.

Habitat
This cactus is found on silty, loamy flats and gravelly slopes among rocks of igneous and limestone substrates associated with creosotebush, lechuguilla, and other desert scrub in oak-juniper woodlands. Rare throughout its range in the Trans-Pecos, from El Paso to Brewster counties, but locally common in Presidio County, at elevations of 800 to 1,900 m (2,500–5,900 ft). Texas Area 10.

Flowering Season
This species blooms from March through April. Flowers open in the morning for 2–3 days.

Notes
Other Common Names: Big Bend eggs, dense mammillaria
Synonyms: *Coryphantha dasyacantha, E. dasyacantha, Mammillaria dasyacantha*
Look-alike Species: *E. dasyacantha* var. *chaffeyii, E. duncanii*

Escobaria emskoetterana — Junior Tom Thumb cactus

Features
Escobaria emskoetterana has many light tan, bristly spines nearly obscuring stems. Diffuse roots give rise to irregular masses of many-branched stems, 10–30 cm across. Ovoid to cylindrical stems, 3–6 cm long and 2.5–3.0 cm in diameter with tubercles 4–8 mm. Round areoles at the tubercle apex age to oval with woolly groove.

Spines
Central: 3–9 porrect, ascending or descending, 1.3–2.3 mm.
Color: Pale yellow or tan, reddish brown to black in distal portion.
Radial: 20–40, some elongated, contorted, 6–10 mm.
Color: Snow white or gray.

Flowers
Nearly apical, greenish yellow with pinkish to brown midstripes, 10–19 mm. Filaments: white; Anthers: yellow; Stigma: dark green to yellow; Style: cream to green.

Fruits
Bright red ellipsoid, cylindrical, or obovoid fruits, 1–2 cm.

Seeds
Black, pitted, spherical seeds, 1.2 mm.

Habitat
This species is found in limestone, gravelly, or silty substrates in dense scrub with lechuguilla from the Pecos River southward to McAllen, Texas, at elevations of 50–400 m (160–1,300 ft). Texas Areas 2, 6, and 7.

Flowering Season
This cactus is expected to bloom in February and March.

Notes
Other Common Names: Big nipple cactus, Runyon's escobaria
Synonyms: *Coryphantha pottsiana, Echinocactus pottsiana, Escobaria robertii, E. runyonii*
Look-alike Species: *Coryphantha sneedii*

Escobaria minima — Nellie's pincushion cactus

Features
Escobaria minima is a tiny, marble-sized cactus with peglike spines. Diffuse or short taproots produce dark green, spherical to short cylindroid stems, 1–2 cm long and 7–17 mm in diameter. Usually unbranched, but 2–3 branches occur infrequently. Tubercles are 2–4 mm diameter with circular areoles, 2 mm.

Spines
Central: None obvious.
Radial: 15–28 peglike, appressed against stem, short, thick, and laterally compressed at the base, 3.5–5.0 mm.
Color: Light tan to gray, sometimes with darker tips.

Flowers
Nearly apical magenta flowers with weak midstripes are as large as the stems, 1.5–2.7 cm. Filaments: greenish; Anthers: bright yellow; Stigma: green; Style: greenish.

Fruits
Green to yellow-tinged, spherical to ovoid fruits, 1.5–6.0 mm.

Seeds
Black, obovoid or pyriform, pitted seeds, 0.8–1.0 mm.

Habitat
This pincushion is found on novaculite ridges and grasslands and is frequently closely associated with Selaginella spikemoss. Unique and endemic to the central Marathon Basin within the Trans-Pecos in Brewster County at elevations of 1,200–1,400 m (4,000–4,600 ft). Texas Area 10.

Flowering Season
This cactus is expected to bloom April through May and opportunistically thereafter following rains. Flowers open midday, close before dark, and do not open again.

Notes
Other Common Names: Bird foot cactus, dwarf cory cactus, Nellie's cory
Synonyms: *Coryphantha minima*, *Mammillaria nellieae*
Look-alike Species: None

Escobaria sneedii var. *orcuttii*   Silverlace cactus

Features
Escobaria sneedii var. *orcuttii* is an erect, mostly solitary plant; older plants are infrequently clustering. White and bristly appearing, cylindrical stems are obscured by spines, 7–25 cm long and 2.5–6.5 cm in diameter. Tubercles are 6–11 mm.

Spines
Central: 10–20 porrect, straight, spreading or appressed with bulbous bases, 9–26 mm.
Color: Snow white, some with dark or cyanic tips.
Radial: 24–35 appressed, 3–14 mm.
Color: Snow white.

Flowers
Nearly apical, pink flowers, 1.5–3.0 cm. Filaments: white; Anthers: yellow; Stigma: white to pink; Style: pink.

Fruits
Red or pale green, cylindrical to obovoid fruits, 1.0–1.7 cm.

Seeds
Reddish brown, pitted, comma-shaped seeds, 1.0–1.5 mm.

Habitat
This species is found on slopes and rocky limestone outcrops in Chihuahuan desert scrub in the southern Trans-Pecos region in Brewster and Presidio counties at elevations of 550–1,350 m (1,800–4,500 ft). Texas Area 10.

Flowering Season
This cactus is expected to bloom in the spring from March through May about noon for 2–3 days.

Notes
Other Common Names: Snowcone nipple cactus, white column, white-spine cob cactus
Synonyms: *Coryphantha albicolumnaria, C. sneedii* var. *albicolumnaria, Mammillaria albicolumnaria*
Look-alike Species: *Mammillaria pottsii*

Escobaria sneedii var. *sneedii* — Sneed's pincushion cactus

Features
Escobaria sneedii var. *sneedii* is cespitose, branching with as many as 100 or more branches in some clumps. Snowy white and bristly appearing, spherical to cylindrical stems are obscured by spines, 3–27 cm long and 1.3–7.0 cm in diameter. Diffuse roots or sometimes short, fleshy taproots. Tubercles are 3.5–12.0 mm.

Spines
Central: 1–5, porrect, straight, radiating or appressed with bulbous bases, 3–15 mm.
Color: Snow white to rarely tan, some with dark tips.
Radial: 25–52 appressed, 3–14 mm.
Color: Snow white.

Flowers
Nearly apical pale pink to whitish flowers, 7–25 mm. Filaments: white to pinkish or magenta; Anthers: yellow; Stigma: white or yellowish white; Style: white.

Fruits
Crimson red or green, cylindrical to obovoid fruits, 5.5–21.0 mm.

Seeds
Bright reddish brown or brownish orange, pitted, comma-shaped seeds, 1.0–1.6 mm.

Habitat
This species is found often on steep, south-facing slopes in rocky limestone outcrops in Chihuahuan desert scrub to conifer woodlands in the western Trans-Pecos region in El Paso, Hudspeth, Presidio, and Brewster counties at elevations of 600–2,600 m (2,000–8,500 ft). Texas Area 10.

Flowering Season
This cactus is expected to bloom in the spring from March through May about noon for 3–14 days.

Notes
Other Common Names: Carpet foxtail cactus, Guadalupe pincushion cactus, Lee's pincushion cactus, silver-lace cactus
Synonyms: *Coryphantha sneedii* var. *sneedii*, *Mammillaria sneedii*
Look-alike Species: *Coryphantha vivipara*

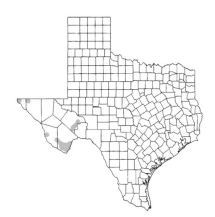

Escobaria tuberculosa — Cob cactus

Features
Escobaria tuberculosa is a solitary cactus or sometimes small-stemmed branching plant with as many as 50 branches. Diffuse roots produce gray-green to green, ovoid to cylindrical stems, 4–16 cm long and 3–6 cm in diameter. Tubercles are firm, 3–6 mm with circular, 2.5–3.0 mm areoles.

Spines
Central: 3–4 straight, porrect, 1.0–1.5 cm.
Color: Pale gray or white with dark tips.
Radial: 15–41 straight, 7–12 mm.
Color: Gray with dark tips.

Flowers
Apical, or nearly so, white to pale pink flowers with inconspicuous midstripes, 2.0–4.5 cm. Filaments: cream; Anthers: pale yellow or white; Stigma: white; Style: white.

Fruits
Bright red ellipsoid, cylindrical, or obovoid fruits, 1.3–2.5 cm.

Seeds
Reddish brown, obovoid, pitted seeds, 0.9–1.0 mm.

Habitat
Cob cactus is found in sedimentary and igneous rocks or novaculite on mountainsides, stony grasslands, and oak-juniper woodlands, often with lechuguilla and creosotebush in the Trans-Pecos area at elevations of 500–2,200 m (1,600–7,200 ft). Texas Areas 7 and 10.

Flowering Season
This species is expected to bloom from April to August in the afternoon for only one day.

Notes
The tubercles of older stems shed their spines as they age and become naked. The bare tubercles develop the appearance of a corn cob, hence the common name.
Other Common Names: Cob cory cactus, corn cob escobaria, varicolor cob cactus, white column foxtail cactus
Synonyms: *Coryphantha tuberculosa*, *Mammillaria tuberculosa*
Look-alike Species: *Coryphantha sneedii*, *E. dasyacantha*

Escobaria vivipara var. *neomexicana* New Mexico beehive cactus

Features
Escobaria vivipara var. *vivipara* appears as a densely spine-covered plant. Diffuse roots support globose to cylindrical stems, 5–20 cm long and 3–8 cm in diameter. Plants are usually solitary, but old plants branch with as many as 6 stems. Tubercles are 8–20 mm and are mostly hidden by the spines.

Spines
Central: 4–10 straight, diffusely radiating or in ascending, appressed "bird-foot" arrangement, 1–20 mm.
Color: Snowy white, tan, or reddish brown.
Radial: 20–40 straight, appressed, glabrous, 8–20 mm.
Color: White, or ashy tan some with dark tips.

Flowers
Subapical, magenta flowers with darker midstripe, 2.5–7.0 cm. Filaments: magenta; Anthers: bright yellow; Stigma: white.

Fruits
Green to brownish red ovoid to obovoid fruits, 1–3 cm.

Seeds
Bright reddish brown, comma-shaped seeds, 1.3–3.0 mm.

Habitat
This beehive cactus is found in rock outcrops, alluvial and sedimentary soils of desert grasslands, and desert scrub to mountain forests. It is found in the northwestern counties of the Trans-Pecos at elevations of 200–2,700 m (620–8,400 ft). Texas Area 10.

Flowering Season
This species is expected to bloom in late spring and summer from May through July in the afternoon, usually for only one day.

Notes
Other Common Names: Fragrant cactus, sour cactus, spiny-star cactus
Synonyms: *Coryphantha neomexicana, C. radiosa* var. *neomexicana, Mammillaria vivipara* var. *neomexicana*
Look-alike Species: *E. hesteri, E. tuberculosa*

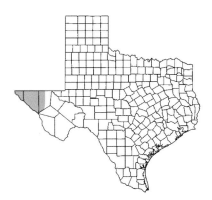

Mammillaria pottsii — Potts' mammillaria

Features
Mammillaria pottsii are recognized as slender plants that branch from the base with stems obscured by whitish spines. Diffuse roots produce blue-green, narrowly cylindrical to clavate stems, 6–20 cm long and 2.0–3.5 cm in diameter, with 3–5 mm tubercles, with abundant white wool found in the axils.

Spines
Central: 6–12 robust, terete, straight, upper central curved upward, not hooked, 5–15 mm.
Color: Pale reddish, tan, gray to white, darker distal portions.
Radial: 37–49 slender, bristlelike, 3–6 mm.
Color: White to pale tan.

Flowers
Small, campanulate flowers are found in a subapical ring around stem and are maroon-red, rusty red, or reddish purple, 6–13 mm. Filaments: pinkish; Anthers: cream; Stigma: reddish purple to orange; Style: reddish.

Fruits
Bright red, cylindrical to clavate fruits, 1.5–2.0 cm.

Seeds
Light to dark brown, nearly oval seeds, 1.0–1.2 mm.

Habitat
This species grows in gravelly flats and rocky slopes of various limestone sediments within Chihuahuan desert scrub and often with lechuguilla. Commonly found in southwestern Big Bend in Presidio and Brewster counties at elevations of 700–2,100 m (2,300–6,900 ft). Texas Areas 7 and 10.

Flowering Season
This cactus is expected to bloom from February to March. Flowers open about midday, close at night, and may reopen for 3–6 days thereafter.

Notes
Other Common Names: Foxtail cactus, Potts' nipple cactus, rat-tail cactus
Synonym: *Coryphantha pottsii*
Look-alike Species: Young, single-stemmed individuals may look like *Sclerocactus mariposensis*.

Neolloydia conoidea   Texas cone cactus

Features
Neolloydia conoidea is a multi-stemmed cactus with prominent tubercles and white woolly apical hairs. Diffuse roots produce gray-green to yellowish green globose to cylindrical stems, clearly visible through the spines, 5–12 cm long and 2.5–7.0 cm in diameter. Plants are usually branched with several stems and very prominent tubercles with felted areolar groove. Areoles are circular, 2–5 mm. Young areoles bear white wool.

Spines
Central: 2–4 straight and stout, terete with a bulbous base, lower central porrect, uppers angled upward, 1.7–2.4 cm.
Color: Lower central blackish, uppers dark to gray with dark tips.
Radial: 14–17 straight, needlelike, and appressed against stem, 6–12 mm.
Color: Ashy white, sometimes with dark tips.

Flowers
Apical magenta, bright rose-pink striped flowers, 3.0–5.5 cm. Filaments: whitish; Anthers: deep yellow; Stigma: white; Style: whitish.

Fruits
Green to greenish brown or tan fruits, spherical, inconspicuous, hidden by apical hair, 4–8 mm.

Seeds
Black to gray, papillate, pyriform seeds, 1.0–1.6 mm.

Habitat
Cone cactus is found in rocky, limestone substrates within desert mountains west of Sanderson, Texas, and west to El Paso County at elevations of 1,500–1,800 m (4,900–5,900 ft). Texas Areas 7 and 10.

Flowering Season
This species is expected to bloom from May through June by late morning, close in the afternoon, and open again for 1–3 days.

Notes
Other Common Names: Cone cactus, Texas cactus
Synonyms: *Echinocactus conoideus, Mammillaria conoidia, N. texensis*
Look-alike Species: *Coryphantha ramillosa, Escobaria macromeris*

Sclerocactus intertextus var. *dasyacanthus*   Longcentral woven-spine pineapple cactus

Features
Sclerocactus intertextus var. *dasyacanthus* is identifiable by its shaggy appearance and frequent dark reddish hue. Diffuse roots produce a reduced, narrowly cylindrical, stalklike stem below the normal aboveground portion. Bristly, green, globose to ovoid to short cylindrical stems are solitary, 7–17 cm long and 2–8 cm in diameter. Tuberculate ribs number 13 with elliptic to nearly circular areoles with felted areolar groove.

Spines
Central: 1 porrect adaxial central, others spreading, 4–15 mm.
Color: Gray or brown to dull red, reddish, or pink-tinged tips.
Radial: 16–24 slightly appressed, 2.4–8.0 cm.
Color: As the centrals.

Flowers
Apical white flowers, 2.3–3.0 cm. Filaments: pale green; Anthers: cream yellow; Stigma: bright red or pink; Style: greenish.

Fruits
Green, globose to ovoid fruits, 8–15 mm.

Seeds
Glossy black, papillate, nearly spherical to reniform seeds, 1 mm.

Habitat
This cactus is found on stony ridges and in igneous substrates (rarely limestone) in plains and desert grasslands on the upper edge of the Chihuahuan Desert in southeastern New Mexico, entering Texas in the Franklin Mountains and in Culberson County. It grows at elevations of 1,100–2,100 m (3,400–6,500 ft). Texas Area 10.

Flowering Season
This species is expected to bloom from late February through early April. It often produces as many as 20 buds simultaneously on a single stem, blooming in succession for as many as 30 days. Individual flowers close at night and reopen for 2–3 days.

Notes
Other Common Names: Early bloomer, white biznagita, white-flowered visnagita
Synonyms: *Echinocactus intertextus* var. *dasyacanthus*, *Echinocactus intertextus* var. *dasyacanthus*
Echinomastus intertextus,
Look-alike Species: *S. warnockii*

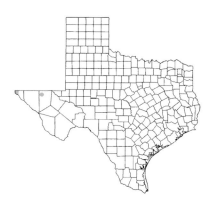

Sclerocactus intertextus var. *intertextus* Woven-spine pinapple cactus

Features
Sclerocactus intertextus var. *intertextus* is identifiable by its smooth appearance, appressed spines, and frequent dark reddish hue. Diffuse roots produce a reduced, narrowly cylindrical, stalklike stem below the normal aboveground portion. Green, spherical to ovoid solitary stems are visible through the spines, 7–20 cm long and 2–8 cm in diameter. Tuberculate ribs number 13, with elliptic to nearly circular areoles with felted areolar grooves.

Spines
Central: 2–3 tightly appressed, 10–19 mm; lower central short, porrect, 1–3 mm.
Color: Gray or brown to dull red, reddish or pink-tinged tips.
Radial: 13–25 tightly appressed, 8–20 mm.
Color: As the centrals.

Flowers
Apical white flowers, 2.3–3.0 cm. Filaments: pale green; Anthers: cream yellow; Stigma: bright red or pink; Style: greenish.

Fruits
Green, globose to ovoid fruits, 8–15 mm.

Seeds
Glossy black, papillate, nearly spherical to reniform seeds, 1 mm.

Habitat
This species is the most common S. intertextus variety. It is abundant in igneous substrates in desert grama grasslands, grassy hills, washes, and plains of the Davis Mountains and northwest into Hudspeth County at elevations of 1,000–1,900 m (3,300–6,200 ft). Texas Area 10.

Flowering Season
This cactus is one of the early-blooming species of the Trans-Pecos, expected to bloom in late February into early April. It often produces as many as 20 buds simultaneously on a single stem. Occasionally, several open at the same time, with the others blooming in succession for as many as 30 days. Individual flowers close at night and reopen for 2–3 days.

Notes
Other Common Names: Early bloomer, white biznagita, white-flowered bisnagita
Synonym: *Echinocactus intertextus*, *Echinomastus intertextus*, *E. intertextus* var. *intertextus*
Look-alike Species: *Sclerocactus warnockii*

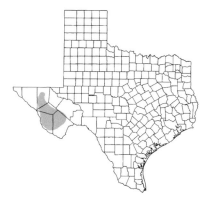

Sclerocactus mariposensis — Mariposa cactus

Features
Sclerocactus mariposensis is a small, ashy white, golf– or tennis ball–size plant with upswept dark blue central spines. Diffuse roots produce solitary stems nearly obscured by overlapping spines. Blue-green, subglobose to cylindrical stems, 3–10 cm long and 3–6 cm in diameter. The cactus may have 21 ribs or may be strictly tuberculate. Areoles along ribs near the plant base produce 20–30 appressed, tiny white spines, 2 mm long.

Spines
Central: 4 porrect or down-curved lower, upper ascending with bulbous bases, 1–2 cm.
Color: Chalky blue or blue-black distally.
Radial: 19–26 slender, horizontally oriented, overlapping, 3–11 mm.
Color: Ashy white.

Flowers
Apical white flowers with pinkish, yellowish tan, or pale green midstripes, 2–3 cm. Filaments: colorless to pink; Anthers: cream to darker yellow; Stigma: green to yellow-green; Style: pale green.

Fruits
Green to yellow-green, globose to oblong fruits, 1 cm.

Seeds
Black, subglobular, papillate seeds, 1.6–2.0 mm.

Habitat
This cactus is found on rocky sites, on barren hills, and in limestone soils in Chihuahuan desert scrub, in southern Presidio and Brewster counties at elevations of 500–1,300 m (1,600–4,300 ft). Texas Area 10.

Flowering Season
This species is an early-blooming species expected to flower only from late February into March. Flowers open midday, close at night, and reopen on warm, sunny days for 3–4 days.

Notes
Other Common Names: Lloyd's mariposa cactus, silver column cactus
Synonyms: *Echinocactus mariposensis, Echinomastus mariposensis, Neolloydia mariposensis*
Look-alike Species: Young specimens look like *Escobaria tuberculosa, Mammillaria pottsii*

Sclerocactus scheeri Root cactus

Features
Sclerocactus scheeri is a globular to cylindrical columnar cactus 17 cm long and 8 cm in diameter. Medium to dark green stems grow from a long, white, fleshy, and tuberous taproot up to 1 m with a fragile constriction at the stem base. The stems branch when injured or when old. Conical or slightly compressed 13 mm tubercles develop into 13 well-defined spiraling or straight tuberculate ribs. Ovate areoles are grooved above and may produce extrafloral nectaries.

Spines
Central: 3–4, one strongly hooked, porrect upper central, pale yellow or brown; two additional straight upper centrals in upright "V." When present, lower centrals straight, erect, and dorsoventrally compressed.
Color: All centrals except the hooked upper central: tan to whitish, sometimes striped brown to black, 1.0–4.5 cm.
Radial: 13–28 slender, 6–28 mm.
Color: Translucent yellow with red-brown tips.

Flowers
Flowers are green to greenish yellow and are produced near stem apex. Only open partially, 1.2–2.0 cm. Filaments: green to reddish; Anthers: yellow to pale orange; Stigma: pale green.

Fruits
Pinkish yellow, club-shaped fruits, 1.9–3.5 cm.

Seeds
Dark brown, compressed globular, 1 mm.

Habitat
This cactus is found in sandy, loamy, silty, or gravelly soils of Tamaulipan thorn scrub, plains, and low hills from northwest of Eagle Pass in Val Verde County southeast along the Rio Grande at elevations of 20–500 m (60–1,600 ft). Texas Areas 6 and 7.

Flowering Season
This species blooms from March through November in the morning and afternoon.

Notes
Other Common Name: Scheer's fishhook cactus
Synonyms: *Ancistrocactus scheeri, Echinocactus scheeri*
Look-alike Species: *Ferocactus hamatacanthus, Glandulicactus uncinatus, S. brevihamatus, Thelocactus setispinus*

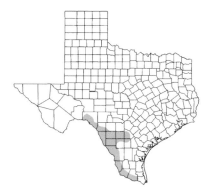

Sclerocactus uncinatus var. *wrightii*   Eagle-claw cactus

Features
Sclerocactus uncinatus var. *wrightii* has abundant, hooked radial spines. Diffuse roots produce solitary stems, occasionally branched at the base. Green, blue-green, to grayish globose, short cylindrical to ovoid stems have a gray waxy appearance, 7–15 cm long and less than 8 cm in diameter. The 9–13 deeply notched prominent ribs have 4.5 mm diameter areoles with gray or yellowish wool in the youngest areoles.

Spines
Central: 1–3 principal central spines prominently hooked, somewhat flattened, erect, 5–9 cm.
Color: Tan, white, or straw colored and pinkish.
Radial: 8–10 appressed, 3 lower centrals subterete and hooked, 2.0–3.5 cm.
Color: Dark red to reddish tan.

Flowers
Apical spiral of brick red, dark red, or maroon, funnelform flowers, 2–3 cm. Filaments: yellow to maroon; Anthers: yellow; Stigma: dull orange.

Fruits
Bright red fruits with white-fringed scales, 1.5–2.5 cm.

Seeds
Black, oblong to obovoid papillate seeds, 1.3–1.5 mm.

Habitat
This species is found on limestone outcrops, igneous substrates, or alluvium in Chihuahuan desert scrub or grasslands, frequently growing in clumps of grama grass. Almost throughout the Trans-Pecos and isolated elsewhere in the Lower Rio Grande Valley at elevations of 100–2,000 m (330–6,200 ft). Texas Areas 2, 6, 7, and 10.

Flowering Season
This cactus is expected to bloom from March through May. Plants produce several flowers that may open simultaneously in the morning or afternoon when stimulated by warm temperatures. Flowers close partially at night and reopen for another 2–3 days.

Notes
Other Common Names: Brown-flowered hedgehog, cat claw cactus, Turk's head
Synonyms: *Ancistrocactus uncinatus* var. *wrightii*, *Echinocactus uncinatus* var. *wrightii*, *Glandulicactus uncinatus* var. *wrightii*
Look-alike Species: Young *Ferocactus hamatacanthus*, *S. brevihamatus*

Sclerocactus warnockii — Warnock's cactus

Features
Sclerocactus warnockii has solitary, blue-green stems not hidden by spreading spines. Diffuse roots produce globose to short cylindrical stems, 4–20 cm long and 3.0–6.5 cm in diameter. Tubercles coalesce into 13 well-defined, low ribs.

Spines
Central: 1–6 divergent like the radials, single lowest porrect or ascending, 1.2–2.5 cm.
Color: Light yellow to pale brown to pale gray with chalky blue tips.
Radial: 11–17 acicular, divergent and overlapping, 8–21 mm.
Color: Light yellow to pale brown to pale gray with chalky blue tips.

Flowers
Near apical, widely opening white to creamy white, rarely pinkish flowers with pale green midstripe, 2.3–3.0 cm. Filaments: pale green to whitish; Anthers: dark yellow; Stigma: light green; Style: light green.

Fruits
Light green, globular fruits, 7–10 mm.

Seeds
Black, subglobular, papillate seeds, 1.6–2.0 mm.

Habitat
Warnock's cactus is found in gravel, gypsum, and limestone substrates on low hills and alluvial flats in dense Chihuahuan desert scrub with lechuguilla, or with grasses and creosotebush. Common in Big Bend and vicinity, including Presidio, Brewster, Jeff Davis, Culberson, and Hudspeth counties at elevations of 500–1,400 m (1,600–4,600 ft). Texas Areas 7 and 10.

Flowering Season
This cactus is first to bloom in winter and spring in its range from February into early March. Plants produce 1–2 flowers simultaneously, opening in early afternoon on warm sunny days, closing at night, and reopening again for 2–4 days. Flower production may continue for several weeks.

Notes
Other Common Names: Warnock biznagita, white flowered cactus
Synonyms: *Echinomastus warnockii, Neolloydia warnockii*
Look-alike Species: *S. intertextus* var. *intertextus*

Thelocactus bicolor var. *bicolor*   Glory of Texas

Features
Thelocactus bicolor var. *bicolor* is an infrequently branching cactus, recognized by its whitish radial spines that look like wood shavings and mostly obscure the stems. Short taproots give rise to erect, green to gray-green, ovoid to cylindrical stems that are not deep-seated in the substrate, 7.5–18.0 cm long and 3.5–11.0 cm in diameter. Well-defined tubercles coalesce to 8–13 rounded ribs, 1.5–2.5 cm wide. Areoles along ribs produce wool in a groove.

Spines
Central: 1–4 terete to angular, porrect, 2–5 cm.
Color: Red and white, aging to ashy white.
Radial: 13–15 terete, straight or slightly recurved, appressed, 1.2–7.5 cm.
Color: Whitish or gray with reddish coloration about midlength of the spine.

Flowers
Stunning, near apical, magenta to rose-pink flowers with dark red centers, 4–8 cm. Filaments: reddish or yellowish; Anthers: yellow; Stigma: reddish to red-orange, pale reddish brown, or yellow; Style: pinkish red.

Fruits
Green to brownish red ovoid to globular fruits, 7–18 mm.

Seeds
Black, obovoid to pyriform, finely papillate seeds, 1.5–2.5 mm.

Habitat
Glory of Texas is infrequently found in sedimentary, igneous, and alluvial substrates among Chihuahuan desert scrub and Tamaulipan thorn scrub. Its range includes the southern tip of Big Bend in Presidio and Brewster counties, as well as in Starr County in the Lower Rio Grande Valley, at elevations of 80–900 m (260–2,800 ft). Texas Areas 6 and 10.

Flowering Season
This cactus is expected to bloom from March through July and again following summer or fall rains. Individual flowers open from midmorning to early afternoon, close at night, and may reopen the next day.

Notes
Other Common Names: Bicolor cactus, Texas pride
Synonyms: *Ferocactus bicolor, T. pottsii*
Look-alike Species: *T. bicolor* var. *flavidispinus*

Thelocactus bicolor var. *flavidispinus*   Marathon Basin thelocactus

Features
Thelocactus bicolor var. *flavidispinus* is a small, unbranched, deep-seated cactus with green, flat-topped, spherical, or short cylindrical stems, 3–9 cm long and 3.5–6.0 cm in diameter. Tubercles are almost completely divided and confluent only at the bases into 13 poorly defined ribs that often spiral around the stem. Mostly circular areoles are closely spaced and produce yellowish spines that almost obscure the stem.

Spines
Central: 1–3 porrect, short, terete, 1.5–2.0 cm.
Color: Yellowish, aging to gray.
Radial: 15–17 needlelike, laterally compressed at the base, recurved toward stem, 1.5–2.0 cm.
Color: Typically yellowish.

Flowers
Near apical, magenta to rose-pink flowers with dark red center, 4–8 cm. Filaments: yellowish or pale red; Anthers: yellow; Stigma: reddish or yellowish toward the tip.

Fruits
Green to brownish red, ovoid to globular fruits, 7–18 mm.

Seeds
Black, obovoid to pyriform, papillate seeds, 1.5–2.5 mm.

Habitat
This cactus is found in novaculite outcrops in semideserts and grasslands of the Marathon Basin in Brewster County at elevations of 1,200–1,300 m (3,700–4,000 ft). Texas Area 10.

Flowering Season
This species is expected to bloom on sunny days from March through May and opportunistically again following summer rains. Individual flowers open midday, close at night, and may reopen the following day.

Notes
Other Common Name: Flat-spined thelocactus
Synonym: *T. flavidispinus*
Look-alike Species: *T. bicolor* var. *bicolor*

Thelocactus setispinus — Twisted rib cactus

Features
Thelocactus setispinus has 13 ribs and a single hooked central spine. Diffuse roots produce solitary stems, infrequently branching at the base. Deep green, hemispheric stems become spherical or cylindrical, 3.6–12.0 cm long and 4.5–12.0 cm in diameter. Narrow tubercles are raised on the 13 slender, spiraling ribs. Areoles are elliptic to ovate with golden, peglike areolar glands and spines not obscuring the stem.

Spines
Central: 1 acicular, porrect, hooked, 1.0–3.8 cm.
Color: Yellowish, turning ashy gray or reddish brown.
Radial: 10–19 acicular, straight or slightly curved toward the stem, 1.2–3.2 cm.
Color: Yellowish, whitish, or reddish brown.

Flowers
Near apical flowers produced in felty, areolar grooves are yellow with red centers, 4–7 cm. Filaments: reddish to pale yellow; Anthers: pale yellow; Stigma: pale yellow; Style: greenish yellow.

Fruits
Bright red, spherical or elongate fruits, 8–10 mm.

Seeds
Black, obovoid, papillate seeds, 1.0–1.4 mm.

Habitat
This cactus is found in heavy soils or limestone substrates in grasslands and in shrublands in mesquite thickets, principally in Central and South Texas and west to Val Verde County at elevations of 0–300 m (0–1,000 ft). Texas Areas 2 and 4–7.

Flowering Season
This species is expected to bloom from April through October. Several flowers may open simultaneously by midday, close at night, and do not open again.

Notes
Other Common Names: Fishhook cactus, hedgehog cactus
Synonyms: *Hamatocactus bicolor, H. setispinus*
Look-alike Species: *Ferocactus hamatacanthus, Sclerocactus uncinatus*

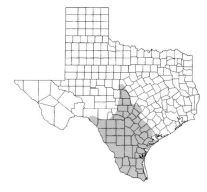

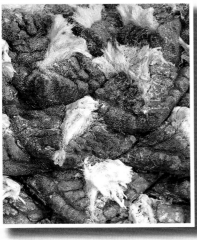

Ariocarpus fissuratus — Living rock cactus

Features
Ariocarpus fissuratus is a flat, deep-seated, solitary cactus with spineless stems. This star-shaped cactus is very cryptic, appearing like the rocks in its habitat. A carrotlike taproot produces gray-green to gray-brown, flattened, and deeply fissured turnip-shaped stems, 5–13 cm. Triangular tubercles are warty, crowded, and overlapping. Elongated areoles are embedded in a 3 mm wide tubercular groove, producing densely matted, grayish wool.

Spines
Central: None.
Radial: None.

Flowers
Pink to magenta funnelform flowers grow from the stem apex; 2.5–5.0 cm. Filaments: white; Anthers: yellow to orange; Stigma: white; Style: white.

Fruits
Oval to cylindrical, light green to white fruits, brown when ripe, 1.0–2.4 cm.

Seeds
Shiny black, globose to obovoid seeds, 1.2–2.5 mm.

Habitat
This cactus is found in limestone chips in rocky hills in the Chihuahuan desert scrub, usually growing with lechuguilla. It ranges from Hudspeth to Val Verde counties at elevations of 500–1,500 m (1,650–5,000 ft). Texas Areas 7 and 10.

Flowering Season
This cactus blooms in the fall from September through November in the morning and afternoon on sunny days for one day only.

Notes
This cactus does not look like a cactus at all. It more closely resembles a flattened, dried flower with leatherlike petals.
Other Common Names: Crack star, dry whiskey, living rock, star cactus
Synonym: *Mammillaria fissurata*
Look-alike Species: None

Astrophytum asterias — Star cactus

Features
Astrophytum asterias is a low, deep-seated, dome-shaped cactus with speckled tufts of hair. Diffuse roots give rise to squat, rounded, shiny green or gray-green stems. The solitary stems are flat-topped and are usually flush with the soil surface, 2.5–6.0 cm long and 5–15 cm in diameter. Usually 8 flat-crested ribs separated by obvious vertical grooves are speckled by bright white hair tufts. Circular, spineless areoles in rows produce dense, dirty yellow or gray wool.

Spines
Central: Absent.
Radial: Absent.

Flowers
Yellow apical flowers with an orange throat open widely, 3.8–5.2 cm. Filaments: yellowish; Anthers: yellow; Stigma: yellow; Style: yellow.

Fruits
Oval to round, green to grayish red fruits are somewhat obscured by white wool, 1–2 cm.

Seeds
Shiny, dark brown to black bowl-shaped seeds, 2–3 mm.

Habitat
Star cactus is found in gravelly soils in Tamaulipan thorn scrub, grasslands, and scrublands and on rocky slopes. It is found only in Star County in the Rio Grande Valley at elevations of 20–100 m (65–350 ft). Texas Area 6.

Flowering Season
This species blooms in March through May and often after rain during the day for one day only.

Notes
This cactus is very cryptic. Its color closely matches that of the surrounding soil, and its low growth makes it difficult to spot.
Other Common Names: Sand dollar, sea urchin cactus, star peyote
Synonym: *Echinocactus asterias*
Look-alike Species: *Lophophora williamsii*

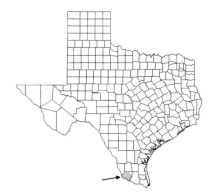

Lophophora williamsii — Peyote

Features
Lophophora williamsii has blue, blue-green, or grayish green, deep-seated stems without obvious spines. Fleshy taproots produce solitary to numerous clustering, depressed globose stems, 2.0–7.5 cm high and 3–10 cm in diameter. The 8 low ribs are marked by grooves delineating low, fused tubercles that produce symmetrically spaced round areoles with tufts of dirty white to yellow wool and no spines.

Spines
Central: Absent.
Radial: Absent.

Flowers
Apical pink to pinkish white flowers with darker pink midstripes, 1.5–2.5 cm. Filaments: white; Anthers: yellow; Stigma: pinkish white; Style: white.

Fruits
White to pinkish, nearly cylindrical and naked fruits partly hidden by apical wool, 1.1–2.5 cm.

Seeds
Black, pear-shaped seeds, 1.0–1.5 mm.

Habitat
Peyote is found in limestone substrates or calcareous alluvium in limestone hills, in crevices, and under protection of other vegetation in Chihuahuan desert scrub and Tamaulipan thorn scrub. While close to being extirpated from Texas, the species' sparse distribution historically was from Big Bend along the Rio Grande into the Lower Rio Grande Valley at elevations of 100–500 m (330–1,500 ft). Texas Areas 6, 7, and 10.

Flowering Season
This species is expected to bloom from March through May.

Notes
Possession of this cactus or any of its parts other than by certain licensed Native Americans is a violation of U.S. federal law. The cactus tissues contain a hallucinogenic alkaloid. For this reason, some Native Americans use the cactus in religious ceremonies.
Other Common Names: Divine cactus, dry whisky, mescal button
Synonym: *Echinocactus williamsii*
Look-alike Species: *Astrophytum asterias*

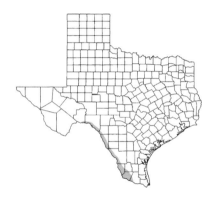

Echinocactus horizonthalonius — Eagle-claw cactus

Features
Echinocactus horizonthalonius is easily identified by its solitary, 8-ribbed subglobose stems with rigid, annulate spines. Diffuse roots or short taproots support gray-green to blue-gray, short cylindrical, hemispheric or flat-topped stems, 4–25 cm long and 8–15 cm in diameter. Plants are solitary and unbranched. Ribs are broadly rounded, vertical, or spiral. Widely spaced areoles are found on the rib crest with gray-white wool on stem apex.

Spines
Central: 1 straight or curved, descending, and 2 appressed, upward curving, 1.8–4.3 cm.
Color: Gray to pink to tan, brownish, or nearly black.
Radial: 5 annulate, ridged, and flattened, curved toward stem, 1.8–4.3 cm.
Color: As the centrals.

Flowers
Rose, pink, or magenta flowers arise from the dense, white wool found on the stem apex, 5–7 cm. Filaments: yellowish; Anthers: yellow; Stigma: pink to olive; Style: pink.

Fruits
Pink or red spherical to ovoid fruits remain hidden by the apical wool, 1–3 cm.

Seeds
Black to gray spherical or obovoid and wrinkled seeds, 2–3 mm.

Habitat
Eagle-claw cactus is found on igneous, sedimentary, clay, and especially limestone and gypsum on arid rocky slopes. It is found throughout the Trans-Pecos and southeastward along the Rio Grande at elevations of 600–1,700 m (2,000–5,600 ft). Texas Areas 6, 7, and 10.

Flowering Season
This cactus is expected to bloom in May and June about midday, and flowers may open for another 2 days.

Notes
Surprising for such an armored cactus, its flesh was used in the past for making cactus candy.
Other Common Names: Bisnaga de dulce, blue barrel cactus, Turk's head cactus, visnaga
Synonyms: None
Look-alike Species: *E. texensis*

Echinocactus texensis Horse-crippler cactus

Features
Echinocactus texensis is a deep-seated cactus with 13 prominent ribs and annulate spines. Diffuse roots produce gray-green to grass green, flat-topped, hemispheric stems, 10–30 cm long and 10–30 cm in diameter. Stem apex is covered with dense, white wool. The 13–27 ribs have very prominent, sharp crests with 3-sided, partially recessed areoles with robust spines and white or gray wool.

Spines
Centrals: 1 straight, descending, 4–6 cm.
Color: Tan, reddish to pink, or gray.
Radials: 6–7 appressed, straight or decurved, 1.4–3.9 cm.
Color: As the centrals.

Flowers
Pink to salmon flowers with prominent red centers develop in the apical wool, 5–6 cm. Filaments: pinkish reddish; Anthers: yellow; Stigma: pink to white; Style: pinkish to yellow.

Fruits
Scarlet-crimson, spherical to ovoid fruits, 1.5–5.0 cm.

Seeds
Glossy black, obovoid, deeply concave seeds, 2.5–3.0 mm.

Habitat
Horse-crippler cactus is found in deep limestone soils, saline flats, and low limestone hills among Chihuahuan desert grasslands, Tamaulipan thorn scrub, and oak woodlands. Distribution is from the eastern Trans-Pecos to the Rio Grande Valley and South Coastal Plains and northward to the Panhandle at elevations of 0–1,400 m (0–4,600 ft). Texas Areas 2–10.

Flowering Season
This species is expected to bloom in late spring from April through May. Flowers open near midday, close at night, and may open for another 3 days.

Notes
This plant's species name, texensis, is appropriate in that the cactus has widespread distribution in Texas and little elsewhere. It is called "horse-crippler" because the robust spines of the frequently hidden cactus inflict damage to the tender flesh under the hoof when stepped on.
Other Common Names: Devil's head, manca caballo
Synonym: *Homalocephala texensis*
Look-alike Species:
E. horizonthalonius

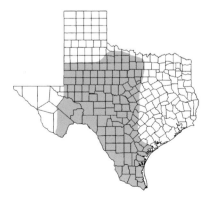

Ferocactus hamatacanthus var. *hamatacanthus*   Giant fishhook cactus

Features
Ferocactus hamatacanthus var. *hamatacanthus* has stem tips with red and yellowish hooked spines. Diffuse roots support dark gray-green, usually solitary or infrequently clustering, hemispheric to cylindrical stems, 10–30 cm tall and 10–25 cm in diameter. The 10–13 thick and large, rounded, and strongly tuberculate ribs are partially obscured by overlapping spines. Areoles are circular to elliptic with white to gray wool in younger areoles.

Spines
Central: 4 angled or terete; 1 lower central hooked, 6–16 cm.
Color: Yellow or stramineus.
Radial: 10–14 terete, upper ascending, lowers descending, laterals appressed, 3.5–7.0 cm.
Color: Gray to reddish, especially near the stem apex.

Flowers
Subapical ring of yellow funnelform flowers, 5.5–8.0 cm. Filaments: yellow to orange-yellow; Anthers: yellow; Stigma: yellow.

Fruits
Green, maturing to dull brownish purple, obovoid to oblong fruits, 3–5 cm.

Seeds
Shiny black, ovate, pitted seeds, 1.5 mm.

Habitat
This species inhabits igneous, calcareous substrates on stony hills, bluffs, and flats in prairies and desert grasslands and under Chihuahuan desert scrub. Wide distribution in southern Trans-Pecos counties west of the Pecos River at elevations of 500–2,000 m (1,600–6,200 ft). Texas Areas 2, 6, 7, and 10.

Flowering Season
This cactus is expected to bloom in the summer from June through August. Flowers open midday, partially close at night, and reopen again for several days.

Notes
Other Common Names: Turk's head, whiskered barrel
Synonym: *F. hamatacanthus* var. *crasispinus*
Look-alike Species: None

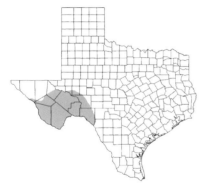

Ferocactus hamatacanthus var. *sinuatus*   Lower Rio Grande Valley barrel cactus

Features
Ferocactus hamatacanthus var. *sinuatus* is a somewhat smaller version of var. *hamatacanthus*. The varieties are geographically separated predominantly east and south of the Pecos River. Variety *sinuatus* is distinguished by smaller stems, 10–30 cm tall and 5–20 cm in diameter. Diffuse roots support dark green to gray-green spherical to ovoid stems. The 13 ribs are narrowly compressed, undulate, and acute at the crest. Areoles are circular to elliptic with white to gray wool in younger areoles.

Spines
Central: 4 terete or somewhat flattened, 1 hooked, 5–9 cm.
Color: Yellow or straw colored.
Radial: 8–12 terete, upper ascending, lower descending, laterals appressed, 3.5–7.0 cm.
Color: Gray to reddish, especially near the stem apex.

Flowers
Apical ring of yellow funnelform flowers, 5.5–9.5 cm. Filaments: yellow to orange-yellow; Anthers: yellow; Stigma: yellow.

Fruits
Green to dark brownish red globose fruits, 2.5 cm.

Seeds
Shiny black, pitted seeds, 1 mm.

Habitat
This species replaces var. *hamatacanthus* in more easterly habitats in Tamaulipan thorn scrub from the Pecos River east and south through Eagle Pass to Brownsville at elevations of 0–500 m (0–1,600 ft). Texas Areas 2, 6, and 7.

Flowering Season
This cactus is expected to bloom in late summer to fall. Like those of var. *hamatacanthus,* flowers open midday, partially close at night, and reopen again for several days.

Notes
Other Common Names: None
Synonym: *Echinocactus sinuatus*
Look-alike Species:
F. hamatacanthus var. *hamatacanthus*

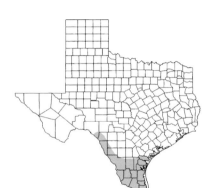

Ferocactus wislizeni — Arizona barrel cactus

Features
Ferocactus wislizeni is the largest barrel cactus in the United States and is recognized for its large, solitary, southward-leaning, heavily ribbed stems. Diffuse roots produce green, aging to brownish, depressed spherical, barrel-shaped, or cylindrical stems, 20–200 cm tall and 19–60 cm in diameter. The 20–30 ribs have a shallow notch above circular to elliptic areoles with white wool and dense spines that partially obscure the stem.

Spines
Central: 4 appear in cross shape, angled outward, lower spines flattened and hooked, all cross-ribbed, 3.6–12.0 cm.
Color: Reddish or grayish.
Radial: 16–20 slender, laterals appressed, 3.5–5.5 cm.
Color: Ashy gray.

Flowers
Subapical ring of showy orange, yellow to red or red-striped flowers, 4.0–8.5 cm. Filaments: yellow to orange-yellow; Anthers: yellow; Stigma: yellow to pale orange-yellow; Style: yellow.

Fruits
Bright yellow, barrel-shaped, scaly fruits, 4–5 cm.

Seeds
Shiny black, finely reticulate seeds, 2.0–2.5 mm.

Habitat
This barrel cactus inhabits deep alluvial and gravelly soils of limestone or igneous origin among desert scrub and grasslands. In Texas this species is at the eastern extreme of its range on southern-facing slopes of the Franklin Mountains in El Paso County at elevations of 1,000–1,800 m (3,300–5,600 ft). Texas Area 10.

Flowering Season
This cactus is expected to bloom from May through June or later following rains. Flowers open near midday, partially close at night, and reopen for several days.

Notes
Other Common Names: Compass barrel, southwestern barrel, visnaga
Synonyms: *Echinocactus emoryi*, *E. wislizeni*
Look-alike Species: Young plants somewhat similar to young *Ferocactus hamatacanthus*

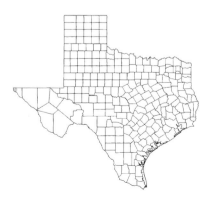

Corypantha echinus — Sea urchin cactus

Features
Coryphantha echinus is easily recognized for its solitary globose stem with the appearance of a white sea urchin and single protruding central spine among the appressed radials. Diffuse or short taproots give rise to unbranched, globose, spherical to cylindrical stems, 3–10 cm long and 3.0–5.5 cm in diameter and mostly obscured by spines. The stems may form large clumps to 80 cm. Tubercles, 8–12 mm, support circular areoles, 2.5–3.0 mm, becoming oval to 4 mm with age, with whitish to brown wool in grooves near the stem apex.

Spines
Central: 1 porrect, straight, or slightly curved with bulbous bases, 1.2–2.5 cm.
Color: Ashy, drab white with brown tips.
Radial: 15–25 appressed, laterally compressed, terete distally, 8–12 mm.
Color: Drab whitish, yellowish, tan or pale gray; overlying bright yellow to dark yellow brown inner layers.

Flowers
Nearly apical, bright yellow, sometimes with a reddish throat, 2.5–6.5 cm. Filaments: red to orange; Anthers: yellow-orange; Stigma: white to greenish yellow; Style: pale yellow.

Fruits
Green, ovoid fruits, 1.2–2.8 cm.

Seeds
Shiny reddish brown, comma-shaped seeds, 1.7–1.9 mm.

Habitat
This *Coryphantha* is found in limestone soils of eroded material. It grows with creosotebush and desert scrub. Endemic in the southeastern Trans-Pecos northward to Howard and Coke counties at elevations of 300–500 m (1,000–1,500 ft). Texas Areas 6–10.

Flowering Season
This species is expected to bloom in late spring and summer as early as April but more frequently from June through July. Flowers open midday, close in early afternoon, and do not open again.

Notes
Other Common Names: Hedgehog cory cactus, rhinoceros cactus
Synonyms: *C. pectinata*, *Mammillaria pectinata*
Look-alike Species: *c. echinus* var. *rovustisina*

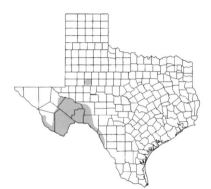

Coryphantha echinus var. *robusta*
Multi-stemmed sea urchin cactus

Features
Coryphantha echinus var. *robusta* has grayish, many-branching stems that form small mounds. Diffuse roots give rise to ovoid or short cylindrical stems, up to 15 cm long and 7 cm in diameter and obscured by dark gray spines. The stems may form mounds to 70 cm.

Spines
Central: 1–4 porrect with bulbous bases, 1.2–2.5 cm.
Color: Gray with black tips.
Radial: 20–30 appressed, 1.2–2.5 cm.
Color: Gray.

Flowers
Nearly apical, bright yellow, 2.5–5.5 cm. Filaments: reddish orange; Anthers: yellow; Stigma: yellow; Style: pale yellow.

Fruits
Green, ovoid fruits, 1.5–2.8 cm.

Seeds
Shiny reddish brown, comma-shaped seeds, 1.7–1.9 mm.

Habitat
This *Coryphantha* is found in limestone or igneous alluvium. It grows with creosotebush and desert scrub or in desert grasslands. It is found in the lower Big Bend area in southern Brewster and Presidio counties at elevations of 700–1,300 m (2,000–3,500 ft). Texas Area 10.

Flowering Season
This species is expected to bloom April through May. Flowers open midday for one day.

Notes
Other Common Names: None
Synonyms: None
Look-alike Species: c. echinus

Coryphantha ramillosa — Whisker brush pincushion cactus

Features
Coryphantha ramillosa is a solitary or sparingly branched cactus, appearing bristly or shaggy gray, brown, or whitish. Plants rarely produce more than 25 branches. Diffuse roots or strong taproot one-fifth of stem diameter. Deep-seated, dark green to gray-green stems are flat-topped or hemispheric, 1–9 cm long and 4–10 cm in diameter. Areolar grooves extend the length of the 8–20 mm tubercle supporting circular areoles, 3 mm. White wool present.

Spines
Central: 1 main central terete, straight, porrect, noticeably projecting; 2–3 others flattened, appressed upward in "bird-foot" pattern, 2.2–4.3 cm.
Color: Dull whitish gray to reddish brown with golden bulbous base.
Radial: 13–16 laterally compressed, sometimes twisted, radiating like spokes, 1.2–3.0 cm.
Color: Glaucous, white.

Flowers
Apical, or nearly so, pale pink to deep rose-purple flowers with darker midstripe, 3.8–6.5 cm. Filaments: white; Anthers: bright yellow to pale orange; Stigma: white; Style: white.

Fruits
Dark to pale gray-green obovoid, spherical, or ellipsoid fruits, 1.6–2.1 cm.

Seeds
Yellow, drying to reddish brown, spherical to comma-shaped seeds, 1.0–1.5 mm.

Habitat
This species is found on limestone ridges and benches in the Chihuahuan desert scrub and frequently found with lechuguilla in Brewster and Terrell counties at elevations of 400–1,000 m (1,300–3,300 ft). Texas Areas 7 and 10.

Flowering Season
This cory cactus is expected to bloom opportunistically from August through November. Flowers open during the morning, close before sunset, and do not open again.

Notes
Other Common Names: Bunched cory cactus, cory cactus
Synonym: *Mammillaria ramillosa*
Look-alike Species: *C. macromeris*

Coryphantha sulcata — Grooved nipple cactus

Features
Coryphantha sulcata appears relatively smooth except for a protruding central spine, partially obscured by spines. Diffuse roots or short taproot. Green, spherical or obovoid stems, 4–8 cm long and 6–8 cm in diameter. Branching, forming clumps (eastern range), sometimes unbranched (western range). Tubercles are soft and flaccid, 8–19 mm. Areoles are 3.5 mm and circular, with white wool. Areolar glands seasonally present.

Spines
Central: 0–4, one straight, porrect; others appressed, 9–15 mm.
Color: Yellowish or pinkish, turning gray to nearly white with darker tips.
Radial: 8–15 stout, 9–16 mm.
Color: Yellowish gray to nearly white with dark tips.

Flowers
Apical flowers, or nearly so, golden yellow with bright red center, 4–6 cm. Filaments: bright red; Anthers: yellow; Stigma: whitish or greenish yellow; Style: yellow.

Fruits
Green becoming dull red, ellipsoid to ovoid fruits, 1.5–2.5 cm.

Seeds
Shiny, dark reddish brown, smooth, comma-shaped seeds, 2 mm.

Habitat
This cactus is found in gravel, sand, or clay in grasslands, scrublands, and savannahs. Generally distributed from the Pecos River eastward through Central Texas, and the Edwards Plateau northward into Oklahoma at elevations of 300–1,100 m (1,000–3,600 ft). Texas Areas 3–8 and 10.

Flowering Season
This species is expected to bloom from April through May for 2–3 days except in hot temperatures.

Notes
Other Common Names: Finger cactus, nipple cactus, pineapple cactus
Synonym: Mammillaria sulcata
Look-alike Species: *C. echinus* (western range), *C. missouriensis* (eastern range)

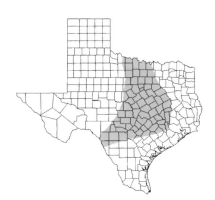

Echinocereus viridiflorus var. *davisii* Dwarf hedgehog cactus

Features
Echinocereus viridiflorus var. *davisii* is a tiny, usually unbranched plant and most cryptic, being deep-seated and nearly covered with soil or other plant matter. It is one of the smallest cacti in the world. Diffuse roots produce erect, dark green, spherical to ovoid stems, 2–3 cm long and 1–2 cm in diameter. Spines partially obscure stems. The 6–9 low ribs are deeply divided into prominent tubercles, 2.0–2.5 mm with elliptic areoles, mostly naked or with a little wool.

Spines
Central: 0–1 terete when present, curving randomly, 1.0–1.2 cm.
Color: Gray or ashy white; brown or reddish tipped.
Radial: 8–15 stout, terete, or slightly flattened and somewhat pectinately arranged, 1.0–2.5 cm.
Color: As the centrals.

Flowers
Bright greenish yellow flowers with maroon midstripes open widely in full sun, frequently larger than the stems, 1.5–2.0 cm. Flowers have an obvious lemon scent. Filaments: pale green; Anthers: yellow; Stigma: green; Style: pale green.

Fruits
Green to reddish brown fruits, 5.5–9.0 cm.

Seeds
Black, tuberculate seeds, 0.9–1.0 mm.

Habitat
This cactus inhabits caballos novaculite substrate in semidesert grasslands, often in mats of Selaginella spikemoss. The species is unique to the Marathon Basin in Brewster County at elevations of 1,200–1,300 m (4,000–4,300 ft). Texas Area 10.

Flowering Season
These plants are expected to bloom in March and April in the morning and afternoon for 3–4 days.

Notes
Other Common Name: Davis hedgehog cactus
Synonym: *E. davisii*
Look-alike Species: *Coryphantha hesteri* (are small but distinguished by spines), *C. minima*

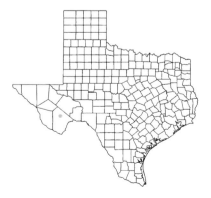

Epithelantha bokei   Boke's button cactus

Features
Epithelantha bokei is a miniature, not deep-seated, flat-topped cactus with a smooth white or yellowish appearance, completely obscured by spines. Diffuse roots produce simple or clustering concave or flat-topped, short cylindrical stems, 1–4 cm long and 2–5 cm in diameter. Tubercles are very small with subcircular areoles and not visible through the spines.

Spines
Central: Not differentiated.
Radial: 30–50 or more, appressed, barely overlapping, 1 mm.
Color: White or cream.

Flowers
Apical, inconspicuous pale pink to silvery white flowers with pinkish or yellow-green light midstripes, 1.0–1.7 cm. Filaments: reddish to pale yellow; Anthers: pinkish to cream; Stigma: white.

Fruits
Green, maturing to bright red, narrowly cylindrical and naked fruits, 8–13 mm.

Seeds
Glossy black, obovoid or comma-shaped seeds, 1.2–1.4 mm.

Habitat
This little cactus is unique to the rocky, barren limestone hills of sedimentary materials in the Chihuahuan Desert in Big Bend and in southern Brewster and Presidio counties at elevations of 700–1,400 m (2,300–4,600 ft). Texas Areas 7 and 10.

Flowering Season
This species is expected to bloom in May through June. Flowers open in the afternoon on warm days, close at night, and reopen again for 1–2 days.

Notes
Other Common Name: Boquillas button cactus
Synonym: *E. micromeris* var. *bokei*
Look-alike Species: *E. micromeris*

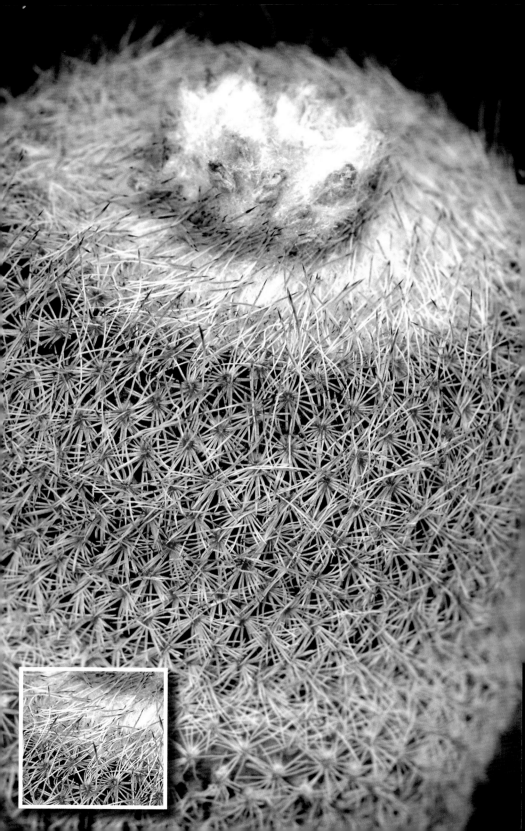

Epithelantha micromeris — Common button cactus

Features
Epithelantha micromeris is a fuzzy white ball that is somewhat rough in appearance. Diffuse roots usually produce solitary spherical stems, 1–4 cm in diameter. Tubercles are tiny, green, and wartlike, with subcircular areoles and many spines.

Spines
Central: Not defined.
Radial: 20–40 slender and straight, slightly overlapping, 2–6 mm.
Color: Whitish to ashy gray with somewhat darker bases.

Flowers
These are the smallest flowers of Texas cacti. Apical, somewhat funnel-form pinkish flowers arise in a tuft of spines, 1.0–1.7 cm. Filaments: reddish to pale yellow; Anthers: pink to cream yellow; Stigma: white.

Fruits
Red, cylindrical, naked fruits, 1–2 cm.

Seeds
Glossy black, comma-shaped seeds, 1.0–1.4 mm.

Habitat
This species inhabits rocky limestone or igneous substrates on hills, ridges in desert, and grasslands. Expected in every Trans-Pecos county at elevations of 500–1,800 m (1,500–5,600 ft). Texas Areas 7 and 10.

Flowering Season
This cactus is expected to bloom from February through April. Several tightly clustered flowers may open in the morning and afternoon on warm, sunny days, close at night, and re-open again for another 1–2 days.

Notes
Other Common Names: Button cactus, mulato, tapon
Synonym: *Mammillaria micromeris*
Look-alike Species: Young *Coryphantha vivipara, E. bokei, Mammillaria lasiacantha*

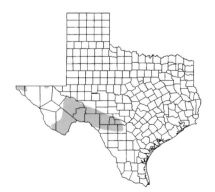

Escobaria duncanii — Duncan's pincushion cactus

Features
Escobaria duncanii is an inconspicuous, difficult-to-identify cactus with shaggy white spines obscuring the stem. The deep-seated, obovoid, or spherical green stem grows from a short, conical taproot about half the diameter of the stem and up to 30 cm long. The stems infrequently branch, are 2.5–6.0 cm long and 1.0–3.4 cm in diameter. Tubercles, 3–6 mm, support brown, circular areoles.

Spines
Central: 0–1 straight when present, 7–14 mm.
Color: Snow white; apical spines may be tan to brown.
Radial: 20–40 slender, 6–9 mm.
Color: White, with light brown bases and barely tipped with brown.

Flowers
Nearly apical, whitish, cream to pink flowers with prominent pinkish to brown midstripes, 1.3–1.9 cm. Filaments: white to pinkish; Anthers: bright yellow; Stigma; green; Style: greenish.

Fruits
Bright red ellipsoid, cylindrical, or clavate fruits, 1.1–2.0 cm.

Seeds
Dull black, pitted, globose seeds, 1.2 mm.

Habitat
Duncan's pincushion cactus is a hard-to-spot species found mostly in crevices on limestone slopes of outcrops in the desert regions of Big Bend in Presidio and Brewster counties at elevations of 700–1,700 m (2,200–5,300 ft). Texas Area 10.

Flowering Season
This cactus is an early bloomer from February through March during warm, sunny days. Flowers close at night and may open again for 2–3 days.

Notes
Other Common Name: Duncan's snowball
Synonyms: *Coryphantha duncanii, E. dasyacantha* var. *duncanii, Mammillaria duncanii*
Look-alike Species: *E. dasyacantha*

Escobaria hesteri — Hester's pincushion cactus

Features
Escobaria hesteri is one of the most cryptic species, often withdrawn below ground and frequently hidden by grasses and other vegetation. Short, fleshy taproots produce deep-seated green to gray-green, solitary, flat-topped or hemispheric stems or sometimes forming multistemmed clumps. Stems are commonly 5–9 cm long and 1.5–4.7 cm in diameter, with circular areoles on tubercles, 4–7 mm. White trichomes are present in young areoles.

Spines
Central: 0–3 acicular, straight when present with tan, bulbous bases, 9–13 mm.
Color: Reddish to brownish gray for half their length.
Radial: 12–23 glabrous, straight, laterally compressed, bulbous bases, 7–13 mm.
Color: White with red or brown tips, weathering to gray.

Flowers
Nearly apical magenta to rose-pink flowers, 1.3–2.0 cm. Filaments: colorless to rose; Anthers: orange to yellow; Stigma: white, cream to pale pink; Style: greenish yellow, reddish above.

Fruits
Green, to greenish red, globose to obovoid fruits, 5–8 mm.

Seeds
Dark brown spherical, pitted seeds, 0.9–1.1 mm.

Habitat
Hester's pincushion cactus is found in areas of rocky crevices of limestone, novaculite, and igneous substrate among *Selaginella* spikemoss, in semidesert grasslands and oak-juniper woodlands. Endemic to the Trans-Pecos and found near Marathon, Texas, in Brewster, Pecos, and Terrell counties at elevations of 1,200–1,600 m (3,700–5,000 ft). Texas Areas 7 and 10.

Flowering Season
This species is expected to bloom from April through June and opportunistically following rain. Flowers open at noon, close in the afternoon, and do not open again.

Notes
Other Common Name: Hester's foxtail cactus
Synonyms: *Coryphantha hesteri*, *Mammillaria hesteri*
Look-alike Species: *E. vivipara*

Escobaria missouriensis — Missouri foxtail cactus

Features
Escobaria missouriensis is a localized and inconspicuous cactus. It has diffuse or short taproots and sometimes roots from the bases of branches. Dark green stems are depressed-globose, 2–8 cm long and 1.8–7.5 cm in diameter and become deep-seated and flat-topped in winter. Frequently forms clumps to 30 cm in diameter. Tubercles, 5–21 mm, with round or oval areoles produce short white wool when young.

Spines
Central: 0–2 porrect to erect or ascending, 9–18 mm.
Color: White, pale gray, or tan, weathering to gray with dark brownish orange to pale gray tips.
Radial: 6–20 appressed, 4–16 mm.
Color: White, gray, or pale tan; dark rusty or gray tips.

Flowers
Nearly apical pale greenish yellow flowers with green, rose-pink, or pale brown midstripes, 2.5–5.0 cm. Filaments: white to light green; Anthers: bright yellow; Stigma: green or yellowish; Style: green to yellow.

Fruits
Green, ripening to orange or scarlet, globose to ellipsoid fruits, 6.5–10.0 mm.

Seeds
Black, deeply pitted, nearly spherical seeds, 1.4–2.2 mm.

Habitat
This species is found among sedimentary rocks and loam in woodlands, plains, and stony, shortgrass prairies with Gambel oak. Its range is somewhat widespread in south-central and North Texas east of the Edwards Plateau from Bexar County northward into Oklahoma and beyond at elevations of 300–2,000 m (1,000–6,300 ft). Texas Areas 1 and 3–9.

Flowering Season
This cactus is expected to bloom from mid-April to June.

Notes
Other Common Names: None
Synonyms:
Coryphantha missouriensis,
Mammillaria missouriensis
Look-alike Species: None

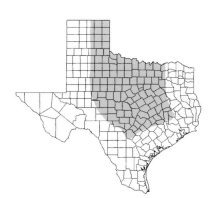

Escobaria vivipara var. *vivipara*   Eastern beehive cactus

Features
Escobaria vivipara var. *vivipara* appears as a sparsely and coarsely spine-covered plant. Diffuse roots, less than one-fourth the stem diameter, support bright green, ovate to spherical stems aging to cylindrical, 2.5–10.0 cm long and 3–11 cm in diameter. Plants are often solitary, but some have as many as 30 branches. Tubercles are 8–25 mm.

Spines
Central: 3–8 straight, radiating in "bird-foot" arrangement, bulbous bases, 9–25 mm.
Color: White, ashy, or reddish brown, darkening with age.
Radial: 10-25 appressed, pectinate, 7–22 mm.
Color: Bright white, ashy, or tan with dark tips.

Flowers
Subapical, pale rose-pink to magenta flowers with darker midstripe. 2.5–7.0 cm. Filaments: magenta; Anthers: bright yellow; Stigma: pink to magenta.

Fruits
Green to brownish red ovoid to obovoid fruits, 1–3 cm.

Seeds
Bright reddish brown, comma-shaped seeds, 1.3–2.4 mm.

Habitat
The most widespread of the Escobaria genera, this beehive cactus is found in alluvial soils of low hills, mountaintops, and desert scrub to conifer forests. It often grows with grass, mesquite, and prickly pear cactus. It is found in most of the Trans-Pecos and eastward through Central Texas and north into Canada at elevations of 200–2,700 m (620-8,400 ft). Texas Areas 4–10.

Flowering Season
This species is expected to bloom in late spring and summer from April through August in the afternoon, usually for only one day.

Notes
Other Common Name: Pincushion cactus
Synonyms: *Coryphantha vivipara*, *Mammillaria vivipara*
Look-alike Species: *E. hesteri*, *E. tuberculosa*

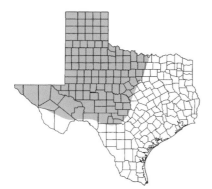

Mammillaria grahamii var. *grahamii* — Graham's fishhook cactus

Features
Mammillaria grahamii var. *grahamii* is a short, globose cactus with a single hooked central spine and 20 or more radials. Roots are diffuse, the upper portion not enlarged. The plants may be solitary or cespitose, branching at the base with 5–10 or more stems. Stems are spherical to cylindrical, 5–16 cm long and 3.5–6.8 cm in diameter. Stems appear whitish to gray due to the covering of radial spines. Ovoid tubercles are firm, 3.5–7.0 mm, with circular areoles.

Spines
Central: 2–3 with one lower porrect, hooked; others erect to appressed, straight, 9.5–25.0 mm.
Color: Reddish to almost black.
Radial: 17–35 bristlelike, 6–12 mm.
Color: Whitish to pale tan.

Flowers
Subapical corona of bright rose-pink or magenta to scarlet flowers, 1.8–3.5 cm. Filaments: pinkish; Anthers: yellow to pale orange; Stigma: yellow-green to green.

Fruits
Green to bright red, clavate fruits, 1.2–2.9 cm.

Seeds
Black, pear-shaped seeds, 0.8–1.9 mm.

Habitat
This fishhook cactus is found in silt, sand, alluvial soils, calcareous gravel, and fields of dark volcanic rock along alluvial slopes, hills, and canyons among Chihuahuan desert scrub, grasslands, and oak woodlands. It is distributed in the Franklin Mountains and northwestern Presidio County at elevations of 80–1,400 m (250–4,400 ft). Texas Area 10.

Flowering Season
This species is expected to bloom as early as May and into June and opportunistically later following rains. Flowers close at night, lasting only one day.

Notes
Other Common Names: Fishhook cactus, lizard catcher
Synonym: *M. microcarpa*
Look-alike Species: *M. wrightii* in the Franklin Mountains

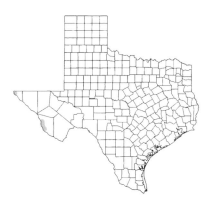

Mammillaria heyderi var. *heyderi* — Heyder's pincushion cactus

Features
Mammillaria heyderi var. *heyderi* is a flattened, solitary, and deep-seated cactus, protruding relatively little above the soil. Large taproots and secondary diffuse roots produce dark green, flat-topped, hemispheric stems, 4–9 cm. Prominent, 3–4 mm tubercles are arranged in spiral rows and are quite visible between the spines. Circular areoles produce white wool, 2 mm. Wool is also found between the tubercles.

Spines
Central: 1, straight and ascending, 2–8 mm.
Color: Brownish with dark tips.
Radial: 13–17 or more, needlelike, 6–11 mm.
Color: Whitish, yellow to brown, often with darker tips.

Flowers
Subapical circle outside new growth of white, greenish, creamy pink flowers with tan, pink, green, or brownish midstripes, 1.9–3.8 cm. Filaments: whitish to pink; Anthers: yellow; Stigma: light green, cream.

Fruits
Bright red to scarlet, obovoid to clavate fruits, 1.0–3.5 cm.

Seeds
Reddish brown, pitted seeds, 1.0–1.2 mm.

Habitat
This cactus is found in gravelly limestone and alluvial substrates under grasses in Chihuahuan desert scrub and Tamaulipan thorn scrub nurse plants. It ranges from El Paso County eastward through the Edwards Plateau and south to Cameron County and the Lower Rio Grande Valley at elevations of 10–1,400 m (33–4,400 ft). Texas Areas 2 and 6–10.

Flowering Season
This cactus is typically expected to bloom in March and April. Flowers open midday, close at night, and re-open again for 3–4 days.

Notes
Other Common Names: Nipple cactus, pancake cactus
Synonyms: *M. heyderi*, *M. heyderi* var. *hemispherica*
Look-alike Species: *M. heyderi* var. *meiacantha*

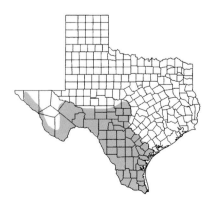

Mammillaria heyderi var. *meiacantha* — Nipple cactus

Features
Mammillaria heyderi var. *meiacantha* is a deep-seated, unbranched cactus with prominent, closely spaced tubercles and relatively thick radial spines. It is the largest *Mammillaria* in the United States. Short, obconic taproots and diffuse secondary roots support deep-seated, dark green to blue-green, flat-topped, hemispheric stems, 10 cm long and 7–16 cm in diameter. Prominent, 8–17 mm sub-pyramidal tubercles produce woolly tufts, 3–5 mm and circular, 2 mm areoles.

Spines
Central: 0–1 porrect or ascending, terete with bulbous base, 5–12 mm.
Color: Reddish brown to nearly black.
Radial: 5–7 needlelike, appressed, and spreading, 6.5–13.0 mm.
Color: Reddish brown to gray with black or dark brown tips.

Flowers
White to pale pink flowers with pink or lavender midstripes develop in ring outside new growth in stem center, 2.5–3.5 cm. Filaments: pinkish white; Anthers: cream yellow; Stigma: green; Style: light green.

Fruits
Purplish pink, clavate to obovoid fruits, 2–3 cm.

Seeds
Pitted, reddish brown seeds, 1.1–1.2 mm.

Habitat
This *mammillaria*, found in desert grasslands and Chihuahuan desert scrub, is widespread in the Trans-Pecos and abundant in southwestern Big Bend at elevations of 900–2,500 m (2,800–8,000 ft). Texas Area 10.

Flowering Season
This species is expected to bloom from May to June when flowers open in late morning to midday for 3–4 days.

Notes
Other Common Names: Little chilies, small-spined pincushion
Synonyms: *M. meiacantha, M. runyonii*
Look-alike Species: *M. heyderi* var. *heyderi*

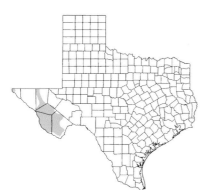

Mammillaria lasiacantha — Golf ball cactus

Features
Mammillaria lasiacantha is a small, inconspicuous, solitary cactus appearing all white, with its stem obscured by spines. It is the first to bloom in the Trans-Pecos. Diffuse roots with upper portions not enlarged support deep-seated, spherical, short cylindrical or depressed stems 2.0–3.5 cm long and 2–4 cm in diameter. Tubercles are 3–6 mm, without wool, and support circular to oval, 1.5 mm areoles.

Spines
Central: Usually lacking.
Radial: 40–60 thin, appressed and overlapping those of adjacent areoles, 2.5–5.0 mm.
Color: White to pale pink, sometimes tipped pinkish brown.

Flowers
Nearly apical flowers are white or cream with well-defined yellow, pink, or reddish brown midstripes, 8–10 mm. Filaments: pale yellow; Anthers: yellow; Stigma: green; Style: greenish.

Fruits
Scarlet, cylindrical or clavate fruits, 1–2 cm.

Seeds
Black, globose or comma-shaped seeds, 1.0–1.2 mm.

Habitat
This species is widespread in the Trans-Pecos and is found among rocky hills and gravelly slopes of predominantly limestone materials within Chihuahuan desert scrub and frequently growing with lechuguilla at elevations of 500–1,800 m (1,500–5,600 ft). Texas Areas 7 and 10.

Flowering Season
This cactus is one of the early bloomers in Texas and is expected to bloom from February through March. Individual flowers open near midday, close at night, and reopen for several days.

Notes
Other Common Names: Fuzzy mammillaria, lace spine cactus
Synonym: *M. denudata*
Look-alike Species: *Epithelantha micromeris*

Mammillaria prolifera var. *texana* — Hair-covered cactus

Features
Mammillaria prolifera var. *texana* forms dense, low mats of many-branching stems with twisting and curved, hairlike radial spines that overlap and nearly obscure the stems. Fibrous roots produce somewhat flaccid stems with as many as 12–20 stems in a single mound. Stems are spherical, cylindrical, or clavate, 3–9 cm long and 1.2–5.0 cm in diameter. Tubercles, 7–10 mm, support circular, 1.5 mm areoles with white wool.

Spines
Central: 5–12 rigid, straight with bulbous bases, spreading, 4–9 mm.
Color: Translucent to pale yellow gradating to reddish brown or black on the distal half.
Radial: 25–60 twisting and curved, overlapping, hairlike, 3–12 mm.
Color: White to pale yellow, dark tipped.

Flowers
Whitish to creamy yellowish, with pinkish to greenish brown midstripes, 1.0–1.8 cm. Filaments: pale yellow to whitish; Anthers: yellow; Stigma: yellowish; Style: cream.

Fruits
Bright red obovoid, clavate, or cylindrical fruits, 8–20 mm.

Seeds
Black, obovate to nearly round seeds, 1.0–1.3 mm.

Habitat
This cactus is found in gravel and limestone crevices in Tamaulipan thorn scrub and oak-juniper woodlands of southwestern Texas and the Rio Grande Valley at elevations of 0–600 m (0–1,800 ft). Texas Areas 2, 6, and 7.

Flowering Season
This species is expected to bloom from March through May.

Notes
Other Common Names: Grape cactus, hair-grooved cactus
Synonyms: *M. multiceps, M. pusilla* var. *texana*
Look-alike Species: None

Mammillaria sphaerica — Pale mammillaria

Features
Mammillaria sphaerica forms low mounds or clumps up to 50 cm high and 30 cm in diameter. Thick, fleshy taproots are enlarged in the upper portions. Stems are light yellow-green, subspherical, 4–5 cm long and 5–8 cm in diameter. Plants may branch from the base, producing as many as 10–30 stems in a single cluster. Loosely spreading tubercles are soft, 1.2–1.5 cm in diameter and 3–5 cm long. Circular, 1.5 mm areoles produce spines not obscuring the stem, and an axial grove produces wool that disappears with age.

Spines
Central: 1 smooth, straight, porrect, and stout, 3–6 mm.
Color: Brownish.
Radial: 12–15 slender, bristlelike with enlarged bases, 6–9 mm.
Color: Yellowish to whitish, aging to gray.

Flowers
Lemon yellow with brownish midlines, 5.5–6.5 cm. Filaments: pale orange; Anthers: pale orange; Stigma: cream to yellow; Style: yellow.

Fruits
Greenish white to dull tan to maroon, ovoid to short cylindrical fruits, 1.0–1.7 cm.

Seeds
Black pitted seeds, 1.3 mm.

Habitat
Pale mammillaria is found in the shade on plains and low gravelly hills in Tamaulipan thorn scrub in South Texas from Laredo to Corpus Christi and southward at elevations of 0–300 m (0–1,000 ft). Texas Areas 2, 6, and 7.

Flowering Season
This species is expected to bloom in July.

Notes
Other Common Name: Long mamma nipple cactus
Synonym: *M. longimamma* var. *sphaerica*
Look-alike Species: *M. longimamma*

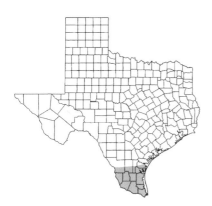

Sclerocactus brevihamatus — Short-spined fishhook cactus

Features
Sclerocactus brevihamatus is a low, deep-seated, and often inconspicuous plant. It is most readily identified by its single, porrect, hooked lower central spine. Dark green to gray-green stems arise from fibrous, diffuse roots or short, conical taproots. Globular unbranched stems are 3–6 cm long and up to 13 cm in diameter. The stems are tuberculate with nearly circular areoles, coalescing into 8–13 ribs in older plants. Whitish wool is frequently present.

Spines
Central: 1–5, lowest central porrect, hooked; others straight, 1.0–4.3 cm.
Color: Gray.
Radial: 7–14 straight, appressed, 1.0–1.8 cm.
Color: Whitish gray.
Wool: Whitish.

Flowers
Inconspicuous funnelform flowers arise from the adaxial edge of areoles at the stem apex. Flowers are pinkish brown, often with reddish brown central stripes, 1–4 cm in diameter. Filaments: rose; Anthers: yellow; Stigma: reddish purple or green to yellow.

Fruits
Pinkish green cylindrical fruits, 8–25 mm.

Seeds
Glossy, dark reddish brown to almost black, kidney-shaped seeds, 1.7–2.0 mm.

Habitat
This cactus is found in sedimentary rocky or gravelly substrates frequently of limestone in Chihuahuan desert scrub, Tamaulipan thorn scrub, oak-juniper woodlands, and poor grasslands. Found in Brewster and Terrell counties and east of the Pecos River to the Edwards Plateau at elevations of 300–1,300 m (900–4,000 ft). Texas Areas 6, 7, and 10.

Flowering Season
This species blooms from January to March in the morning and afternoon. The flowers may open again on the second day.

Notes
Other Common Names: Snipe cactus, and incorrectly, Tobush fishhook cactus
Synonyms: *Ancistrocactus brevihamatus, Echinocactus scheeri* var. *brevihamatus*
Look-alike Species: *Ferocactus hamatacanthus, Glandulicactus uncinatus, Hamatocactus bicolor*

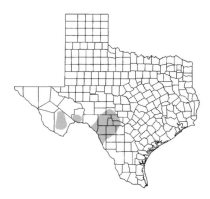

Sclerocactus brevihamatus var. *tobuschii* — Tobusch fishhook cactus

Features
Sclerocactus brevihamatus var. *tobuschii* is a low, deep-seated, and often inconspicuous plant. It is most readily identified by its single, hooked lower central spine. Dark green to gray-green stems arise from fibrous, diffuse roots or short, conical taproots. Globular, unbranched stems, 2–6 cm long and up to 8 cm in diameter. The stems are tuberculate with oval areoles, coalescing into 8–13 ribs in older plants. Tubercular grooves that are half the length of the tubercle frequently have whitish wool.

Spines
Central: 1–5, lowest central gray, porrect, hooked; others straight, 1.0–4.3 cm.
Color: Whitish gray.
Radial: 7–14 straight, appressed, 1.0–1.8 cm.
Color: As the centrals.

Flowers
Inconspicuous funnelform flowers arise from areoles at the stem apex. Flowers are green, 1–4 cm in diameter. Filaments: green; Anthers: yellow; Style: greenish yellow; Stigma: green to yellow.

Fruits
Pinkish green, cylindrical fruits, 8–25 mm.

Seeds
Glossy, dark reddish brown to almost black, kidney shaped seeds, 1.7–2.0 mm.

Habitat
This cactus is found in the Texas Hill Country in thin soils on limestone hilltops and plateaus in oak-juniper woodlands at elevations of 400–500 m (1,200–1,500 ft). Texas Area 7.

Flowering Season
Tobusch fishhook cactus blooms from February through March in the morning and afternoon. Opening of individual flowers may occur for several days.

Notes
Other Common Names: None
Synonyms: *Ancistrocactus tobuschii, Mammillaria tobuschii*
Look-alike Species: *S. brevihamatus, S. scheeri*

The threatened Chisos Mountain hedgehog cactus (*Echinocereus chisosensis*) grows in clumped dog cholla (*Corynopuntia aggeria*) as well as other desert species, which act as nurse plants that provide some protection from the sun and increased moisture and sheltered habitat for the seedling.

Appendix
Species Synonyms

Cacti of Texas—Taxonomy Authority

Classification of the taxonomic relationships of cacti within Texas and, indeed elsewhere, is a formidable task. Many researchers have developed classification models only to have them become obsolete within a few years by the science of chromosome and DNA studies. Although this classification dynamics may continue for some time in the future, the listing of taxa must be reduced to writing by using the best information available at the time of publication of the manuscript.

Much consultation and concern went into this text in the assurance that the taxonomy and naming of species will be as authoritative as possible. It is the desire of the writers to make this guidebook as long lasting and accurate as possible. Many references were used in establishing the classification of cacti within *Texas Cacti*. However, one reference was selected as our choice as the authority at the time of this writing: *The New Cactus Lexicon*, by David Hunt, published in 2006. The naming of species in this text follows that reference.

Other references found useful include the comprehensive work produced by A. Michael Powell and James Weedin in their *Cacti of the Trans-Pecos & Adjacent Areas*; the highly regarded *Flora of North America North of Mexico*, Vol. 4, by the Flora of North America Editorial Committee; and *The Cactus Family*, by Edward Anderson, as well as several older classic references listed in the bibliography.

This decision has created several significant movements of individual species from one genus to another. Some include plants familiar to many of us over a long time. To help summarize these changes, we have included the following table of synonyms using the most common versions found. It is our hope that readers consult this listing to identify the new names and to help find their favorite cactus within the text.

Synonyms
Bold = Illustrated in this Text

Acanthocereus pentagonus
   Syn = *Acanthocereus tetragonus*
   Syn = *Cereus pentagonus*

Acanthocereus tetragonus
   Syn = *Acanthocereus pentagonus*
   Syn = *Cereus pentagonus*

Ancistrocactus scheeri
   Syn = *Echinocactus scheeri*
   Syn = *Sclerocactus scheeri*

Ancistrocactus tobuschii
   Syn = *Sclerocactus brevihamatus* var. *tobuschii*
   Syn = *Sclerocactus tobuschii*

Ancistrocactus uncinatus var. *wrightii*
   Syn = *Echinocactus uncinatus* var. *wrightii*
   Syn = *Glandulicactus uncinatus* var. *wrightii*
   Syn = *Sclerocactus uncinatus* var. *wrightii*

Ariocarpus fissuratus
   Syn = *Mammillaria fissurata*
Astrophytum asterias
   Syn = *Echinocactus asterias*
Cactus pusillus
   Syn = *Opuntia drummondii*
   Syn = *Opuntia pusilla*
Cactus strictus
   Syn = *Opuntia stricta*
   Syn = *Opuntia stricta* var. *dillenii*
Cereus berlandieri
   Syn = *Echinocereus berlandieri*
Cereus fendleri
   Syn = *Echinocereus fendleri*
Cereus greggii
   Syn = *Peniocereus greggii*
Cereus pentagonus
   Syn = *Acanthocereus pentagonus*
   Syn = *Acanthocereus tetragonus*
Cereus pentalophus
   Syn = *Echinocereus pentalophus*
Cereus poselgeri
   Syn = *Echinocereus poselgeri*
Cereus pottsii
   Syn = *Cereus greggii*
   Syn = *Peniocereus greggii*
Cereus stramineus
   Syn = *Echinocereus enneacanthus* var. *stramineus*
   Syn = *Echinocereus stramineus* var. *stramineus*
Corynopuntia aggeria
   Syn = *Grusonia aggeria*
   Syn = *Opuntia aggeria*
Corynopuntia emoryi
   Syn = *Grusonia emoryi*
   Syn = *Opuntia emoryi*
Corynopuntia grahamii
   Syn = *Grusonia grahamii*
   Syn = *Opuntia grahamii*
Corynopuntia schottii
   Syn = *Grusonia schottii*
   Syn = *Opuntia schottii*
Coryphantha albicolumnaria
   Syn = *Coryphantha sneedii* var. *albicolumnaria*
   Syn = *Escobaria sneedii* var. *orcuttii*
   Syn = *Mammillaria albicolumnaria*
Coryphantha chaffeyi
   Syn = *Escobaria dasyacantha* var. *chaffeyi*
Coryphantha dasyacantha
   Syn = *Escobaria dasyacantha* var. *dasyacantha*
   Syn = *Mammillaria dasyacantha*
Coryphantha dasyacantha var. *varicolor*
   Syn = *Escobaria tuberculosa* var. *varicolor*
Coryphantha duncanii
   Syn = *Escobaria duncanii*
Coryphantha echinus
   Syn = *Coryphantha pectinata*
Coryphantha echinus var. **robusta**
Coryphantha hesteri
   Syn = *Escobaria hesteri*
Coryphantha macromeris
   Syn = *Echinocactus macromeris*
   Syn = *Mammillaria macromeris*
Coryphantha macromeris var. **runyonii**
   Syn = *Coryphantha macromeris*
   Syn = *Mammillaria macromeris*
Coryphantha minima
   Syn = *Escobaria minima*
Coryphantha missouriensis
   Syn = *Escobaria missouriensis*
   Syn = *Mammillaria missouriensis*
Coryphantha neomexicana
   Syn = *Coryphantha radiosa* var. *neomexicana*
   Syn = *Escobaria vivipara* var. *neomexicana*
   Syn = *Mammillaria vivipara* var. *neomexicana*
Coryphantha pectinata
   Syn = *Coryphantha echinus*
   Syn = *Mammillaria echinus*
Coryphantha pottsiana
   Syn = *Coryphantha robertii*

Appendix

   Syn = *Echinocactus pottsiana*
   **Syn = *Escobaria
      emskoetterana***
   Syn = *Escobaria runyonii*
Coryphantha pottsii
   Syn = *Mammillaria pottsii*
Coryphantha radiosa var.
neomexicana
   Syn = *Coryphantha neomexicana*
   **Syn = *Escobaria vivipara* var.
      *neomexicana***
   Syn = *Mammillaria vivipara* var.
      neomexicana
Coryphantha ramillosa
   Syn = *Mammillaria ramillosa*
Coryphantha robustispina var.
scheeri
   Syn = *Coryphantha scheeri*
   Syn = *Mammillaria robustispina*
Coryphantha runyonii
   **Syn = *Coryphantha macromeris*
      var. *runyonii***
Coryphantha scheeri var. *scheeri*
   **Syn = *Coryphantha robustispina*
      var. *scheeri***
   Syn = *Mammillaria robustispina*
Coryphantha sneedii
   Syn = *Coryphantha
      albicolumnaria*
   **Syn = *Escobaria sneedii* var.
      *sneedii***
Coryphantha sneedii var.
albicolumnaria
   Syn = *Coryphantha
      albicolumnaria*
   **Syn = *Escobaria sneedii* var.
      *orcuttii***
   Syn = *Mammillaria
      albicolumnaria*
Coryphantha sulcata
   Syn = *Mammillaria sulcata*
Coryphantha tuberculosa
   Syn = *Escobaria tuberculosa*
   Syn = *Mammillaria tuberculosa*
Coryphantha vivipara var.
neomexicana

   **Syn = *Escobaria vivipara* var.
      *neomexicana***
   Syn = *Mammillaria vivipara* var.
      neomexicana
Coryphantha vivipara var. *vivipara*
   **Syn = *Escobaria vivipara* var.
      *vivipara***
   Syn = *Mammillaria vivipara*
Cylindropuntia davisii
   Syn = *Opuntia davisii*
   Syn = *Opuntia tunicata* var.
      davisii
Cylindropuntia imbricata var.
arborescens
   Syn = *Opuntia imbricata* var.
      arborescens
Cylindropuntia imbricata var.
argentea
   Syn = *Opuntia imbricata* var.
      argentea
Cylindropuntia kleiniae
   Syn = *Opuntia kleiniae*
   Syn = *Opuntia wrightii*
Cylindropuntia leptocaulis
   Syn = *Opuntia leptocaulis*
Cylindropuntia tunicata
   Syn = *Opuntia stapeliae*
   Syn = *Opuntia tunicata*
Cylindropuntia wrightii
   Syn = *Cylindropuntia kleiniae*
   Syn = *Opuntia kleiniae*
Echinocactus asterias
   Syn = *Astrophytum asterias*
Echinocactus brevihamatus
   Syn = *Sclerocactus brevihamatus*
Echinocactus emoryi
   Syn = *Echinocactus wislizeni*
   Syn = *Ferocactus wislizeni*
Echinocactus horizonthalonius
Echinocactus intertextus
   Syn = *Echinomastus intertextus*
      var. *intertextus*
   **Syn = *Sclerocactus intertextus*
      var. *intertextus***
Echinocactus intertextus var.
dasyacanthus

Syn = *Echinomastus intertextus*
   var. *dasyacanthus*
   Syn = Sclerocactus intertextus
   var. *dasyacanthus*
Echinocactus pectinatus
   Syn = Echinocereus pectinatus
   var. *wenigeri*
Echinocactus reichenbachii
   Syn = Echinocereus reichenbachii
Echinocactus scheeri
   Syn = *Ancistrocactus scheeri*
   Syn = Sclerocactus scheeri
Echinocactus sinuatus
   Syn = Ferocactus hamatacanthus
   var. *sinuatus*
Echinocactus texensis
   Syn = *Homalocephala texensis*
Echinocactus uncinatus var.
wrightii
   Syn = *Ancistrocactus uncinatus*
      var. *wrightii*
   Syn = *Glandulicactus uncinatus*
      var. *wrightii*
   Syn = Sclerocactus uncinatus
   var. *wrightii*
Echinocactus williamsii
   Syn = Lophophora williamsii
Echinocactus wislizeni
   Syn = Ferocactus wislizeni
Echinocereus berlandieri
   Syn = *Cereus berlandieri*
Echinocereus chisosensis
   Syn = *Echinocereus reichenbachii*
      var. *chisosensis*
Echinocereus chloranthus
   Syn = *Echinocereus chloranthus*
      var. *chloranthus*
   Syn = Echinocereus viridiflorus
   var. *chloranthus*
Echinocereus chloranthus var.
chloranthus
   Syn = *Echinocereus chloranthus*
   Syn = Echinocereus viridiflorus
   var. *chloranthus*
Echinocereus chloranthus var.
cylindricus
   Syn = *Echinocereus stanleyi*
   Syn = Echinocereus viridiflorus
   var. *cylindricus*
Echinocereus coccineus
   Syn = Echinocereus coccineus var.
aggregatus
   Syn = *Echinocereus triglochidiatus*
      var. *paucispinus*
Echinocereus coccineus var.
   aggregatus
   Syn = Echinocereus coccineus
Echinocereus dasyacanthus
   Syn = *Echinocereus pectinatus*
      var. *dasyacanthus*
Echinocereus davisii
   Syn = Echinocereus viridiflorus
   var. *davisii*
Echinocereus dubius
   Syn = Echinocereus enneacanthus
   var. *enneacanthus*
Echinocereus enneacanthus forma
brevispinus
   Syn = Echinocereus enneacanthus
   var. *brevispinus*
Echinocereus enneacanthus var.
brevispinus
   Syn = *Echinocereus enneacanthus*
      forma *brevispinus*
Echinocereus enneacanthus var.
enneacanthus
   Syn = *Echinocereus dubius*
Echinocereus enneacanthus var.
stramineus
   Syn = *Cereus stramineus*
   Syn = Echinocereus stramineus
   var. *stramineus*
Echinocereus fendleri var.
rectispinus
   Syn = Echinocereus fendleri
Echinocereus fendleri
   Syn = *Cereus fendleri*
   Syn = *Echinocereus fendleri* var.
      rectispinus

Echinocereus lloydii
   Syn = *Echinocereus x roetteri*
   **Syn = *Echinocereus x roetteri*
      var. *neomexicana***
Echinocereus pectinatus var. *dasyacanthus*
   Syn = *Echinocereus dasyacanthus*
Echinocereus pectinatus var. *wenigeri*
   Syn = *Echinocactus pectinatus*
Echinocereus pentalophus
   Syn = *Cereus pentalophus*
Echinocereus poselgeri
   Syn = *Cereus poselgeri*
   Syn = *Wilcoxia poselgeri*
Echinocereus reichenbachii
   Syn = *Echinocactus reichenbachii*
Echinocereus reichenbachii var. *chisosensis*
   Syn = *Echinocereus chisosensis*
Echinocereus stanleyi
   Syn = *Echinocereus chloranthus* var. *cylindricus*
   Syn = *Echinocereus viridiflorus* var. *cylindricus*
Echinocereus stramineus var. *stramineus*
   Syn = *Cereus stramineus*
   Syn = *Echinocereus enneacanthus* var. *stramineus*
Echinocereus viridiflorus var. *canus*
   Syn = *Echinocereus viridiflorus* var. *nova*
Echinocereus viridiflorus var. *chloranthus*
   Syn = *Echinocereus chloranthus*
   Syn = *Echinocereus chloranthus* var. *chloranthus*
Echinocereus viridiflorus var. *correllii*
Echinocereus viridiflorus var. *cylindricus*
   Syn = *Echinocereus chloranthus* var. *cylindricus*
   Syn = *Echinocereus stanleyi*

Echinocereus viridiflorus var. *davisii*
   Syn = *Echinocereus davisii*
Echinocereus viridiflorus var. *nova*
   Syn = *Echinocereus viridiflorus* var. *canus*
Echinocereus viridiflorus var. *viridiflorus*
   Syn = *Echinocereus chloranthus* var. *cylindricus*
Echinocereus x roetteri
   Syn = *Echinocereus chloranthus* var. *cylindricus*
   Syn = *Echinocereus x roetteri* var. *neomexicana*
Echinocereus x roetteri var. *neomexicana*
   Syn = *Echinocereus lloydii*
   Syn = *Echinocereus x roetteri*
Echinomastus intertextus var. *dasyacanthus*
   Syn = *Echinocactus intertextus* var. *dasyacanthus*
   Syn = *Sclerocactus intertextus* var. *dasyacanthus*
Echinomastus intertextus var. *intertextus*
   Syn = *Echinocactus intertextus*
   Syn = *Sclerocactus intertextus* var. *intertextus*
Echinomastus mariposensis
   Syn = *Neolloydia mariposensis*
   Syn = *Sclerocactus mariposensis*
Echinomastus warnockii
   Syn = *Neolloydia warnockii*
   Syn = *Sclerocactus warnockii*
Epithelantha bokei
   Syn = *Epithelantha micromeris* var. *bokei*
Epithelantha micromeris
   Syn = *Mammillaria micromeris*
Epithelantha micromeris var. *bokei*
   Syn = *Epithelantha bokei*
Escobaria chaffeyi
   Syn = *Coryphantha chaffeyi*
   Syn = *Escobaria dasyacantha* var. *chaffeyi*

Escobaria dasyacantha* var. *chaffeyi
    Syn = *Coryphantha chaffeyi*
    Syn = *Escobaria chaffeyi*
Escobaria dasyacantha* var. *dasyacantha
    Syn = *Coryphantha dasyacantha*
    Syn = *Escobaria dasyacantha*
    Syn = *Mammillaria dasyacantha*
Escobaria duncanii
    Syn = *Coryphantha duncanii*
    Syn = *Escobaria dasyacantha* var. *duncanii*
    Syn = *Mammillaria duncanii*
Escobaria emskoetterana
    Syn = *Coryphantha pottsiana*
    Syn = *Echinocactus pottsiana*
    Syn = *Escobaria runyonii*
Escobaria hesteri
    Syn = *Coryphantha hesteri*
    Syn = *Mammillaria hesteri*
Escobaria minima
    Syn = *Coryphantha minima*
    Syn = *Mammillaria nellieae*
Escobaria missouriensis
    Syn = *Coryphantha missouriensis*
    Syn = *Mammillaria missouriensis*
Escobaria runyonii
    Syn = *Coryphantha pottsiana*
    Syn = *Coryphantha robertii*
    Syn = *Escobaria emskoetterana*
Escobaria sneedii* var. *orcuttii
    Syn = *Coryphantha albicolumnaria*
    Syn = *Coryphantha sneedii* var. *albicolumnaria*
    Syn = *Mammillaria albicolumnaria*
Escobaria sneedii* var. *sneedii
    Syn = *Coryphantha sneedii* var. *sneedii*
    Syn = *Mammillaria sneedii*
Escobaria tuberculosa
    Syn = *Coryphantha tuberculosa*
    Syn = *Mammillaria tuberculosa*

Escobaria vivipara* var. *neomexicana
    Syn = *Coryphantha vivipara* var. *neomexicana*
    Syn = *Mammillaria vivipara* var. *neomexicana*
Escobaria vivipara* var. *vivipara
    Syn = *Coryphantha vivipara* var. *vivipara*
    Syn = *Mammillaria vivipara*
Ferocactus hamatacanthus var. *crassispinus*
    Syn = *Ferocactus hamatacanthus* var. *hamatacanthus*
Ferocactus hamatacanthus* var. *hamatacanthus
    Syn = *Ferocactus hamatacanthus* var. *crassispinus*
Ferocactus hamatacanthus* var. *sinuatus
    Syn = *Echinocactus sinuatus*
Ferocactus wislizeni
    Syn = *Echinocactus wislizeni*
    Syn = *Echinocactus emoryi*
Glandulicactus uncinatus var. *wrightii*
    Syn = *Ancistrocactus uncinatus* var. *wrightii*
    Syn = *Echinocactus uncinatus* var. *wrightii*
    Syn = *Sclerocactus uncinatus* var. *wrightii*
Grusonia aggeria
    Syn = *Corynopuntia aggeria*
    Syn = *Opuntia aggeria*
Grusonia emoryi
    Syn = *Corynopuntia emoryi*
    Syn = *Opuntia emoryi*
Grusonia grahamii
    Syn = *Corynopuntia grahamii*
    Syn = *Opuntia grahamii*
Grusonia schottii
    Syn = *Corynopuntia schottii*
    Syn = *Opuntia schottii*
Hamatocactus bicolor
    Syn = *Hamatocactus setispinus*
    Syn = *Thelocactus setispinus*

Appendix

Hamatocactus setispinus
   Syn = *Hamatocactus bicolor*
   Syn = Thelocactus setispinus
Homalocephala texensis
   Syn = Echinocactus texensis
Lophophora williamsii
   Syn = *Echinocactus williamsii*
Mammillaria albicolumnaria
   Syn = *Coryphantha albicolumnaria*
   Syn = *Coryphantha sneedii*
   Syn = *Coryphantha sneedii* var. *albicolumnaria*
   Syn = Escobaria sneedii var. orcuttii
Mammillaria conoidia
   Syn = *Echinocactus conoideus*
   Syn = Neolloydia conoidea
   Syn = *Neolloydia texensis*
Mammillaria dasyacantha
   Syn = *Coryphantha dasyacantha*
   Syn = Escobaria dasyacantha var. dasyacantha
Mammillaria denudata
   Syn = Mammillaria lasiacantha
Mammillaria duncanii
   Syn = *Coryphantha duncanii*
   Syn = *Escobaria dasyacantha* var. *duncanii*
   Syn = Escobaria duncanii
Mammillaria echinus
   Syn = Coryphantha echinus
   Syn = *Coryphantha pectinata*
Mammillaria fissurata
   Syn = Ariocarpus fissuratus
Mammillaria grahamii var. *grahamii*
   Syn = *Mammillaria grahamii* var. *oliviae*
   Syn = *Mammillaria microcarpa*
Mammillaria grahamii var. *oliviae*
   Syn = Mammillaria grahamii var. grahamii
Mammillaria heyderi var. *heyderi*
   Syn = *Mammillaria heyderi*
   Syn = *Mammillaria heyderi* var. *hemispherica*

Mammillaria heyderi var. *meiacantha*
   Syn = *Mammillaria meiacantha*
   Syn = *Mammillaria runyonii*
Mammillaria lasiacantha
   Syn = *Mammillaria denudata*
Mammillaria longimamma
   Syn = Mammillaria sphaerica
Mammillaria macromeris
   Syn = Coryphantha macromeris
   Syn = *Escobaria macromeris*
Mammillaria meiacantha
   Syn = Mammillaria heyderi var. meiacantha
   Syn = *Mammillaria runyonii*
Mammillaria micromeris
   Syn = Epithelantha micromeris
Mammillaria missouriensis
   Syn = *Coryphantha missouriensis*
   Syn = Escobaria missouriensis
Mammillaria multiceps
   Syn = Mammillaria prolifera var. texana
   Syn = *Mammillaria pusilla* var. *texana*
Mammillaria nellieae
   Syn = *Coryphantha minima*
   Syn = Escobaria minima
Mammillaria pottsii
   Syn = *Coryphantha pottsii*
Mammillaria prolifera var. *texana*
   Syn = *Mammillaria multiceps*
   Syn = *Mammillaria pusilla* var. *texana*
Mammillaria pusilla var. *texana*
   Syn = *Mammillaria multiceps*
   Syn = Mammillaria prolifera var. texana
Mammillaria ramillosa
   Syn = Coryphantha ramillosa
Mammillaria robustispina
   Syn = Coryphantha robustispina var. scheeri
Mammillaria runyonii
   Syn = Mammillaria heyderi var. meiacantha
   Syn = *Mammillaria meiacantha*

Mammillaria sphaerica
    Syn = *Mammillaria longimamma* var. *sphaerica*
Mammillaria sulcata
    Syn = *Coryphantha sulcata*
Mammillaria tuberculosa
    Syn = *Coryphantha tuberculosa*
    Syn = *Escobaria tuberculosa*
Mammillaria vivipara
    Syn = *Coryphantha vivipara* var. *vivipara*
    Syn = *Escobaria vivipara* var. *vivipara*
Mammillaria vivipara var. *neomexicana*
    Syn = *Coryphantha vivipara* var. *neomexicana*
    Syn = *Escobaria vivipara* var. *neomexicana*
Neolloydia conoidea
    Syn = *Echinocactus conoideus*
    Syn = *Mammillaria conoidia*
    Syn = *Neolloydia texensis*
Neolloydia mariposensis
    Syn = *Echinocactus mariposensis*
    Syn = *Echinomastus mariposensis*
    Syn = *Sclerocactus mariposensis*
Neolloydia texensis
    Syn = *Echinocactus conoideus*
    Syn = *Mammillaria conoides*
    Syn = *Neolloydia conoidea*
Neolloydia warnockii
    Syn = *Echinomastus warnockii*
    Syn = *Sclerocactus warnockii*
Opuntia aggeria
    Syn = *Corynopuntia aggeria*
    Syn = *Grusonia aggeria*
Opuntia atrispina
Opuntia aureispina
    Syn = *Opuntia azurea* var. *aureispina*
    Syn = *Opuntia macrocentra* var. *aureispina*
Opuntia austrina
    Syn = *Opuntia compressa* var. *austrina*
    Syn = *Opuntia humifusa*

Opuntia azurea
    Syn = *Opuntia azurea* var. *diplopurpurea*
    Syn = *Opuntia azurea* var. *parva*
    Syn = *Opuntia violacea* var. *macrocentra*
Opuntia azurea var. *aureispina*
    Syn = *Opuntia phaeacantha*
Opuntia azurea var. *diplopurpurea*
    Syn = *Opuntia azurea*
    Syn = *Opuntia violacea* var. *macrocentra*
Opuntia azurea var. *discolor*
Opuntia azurea var. *parva*
    Syn = *Opuntia azurea*
    Syn = *Opuntia violacea* var. *macrocentra*
Opuntia camanchica
    Syn = *Opuntia phaeacantha* var. *camanchica*
    Syn = *Opuntia phaeacantha* var. *major*
Opuntia chisosensis
    Syn = *Opuntia lindheimeri* var. *chisosensis*
Opuntia chlorotica
    Syn = *Opuntia palmeri*
Opuntia compressa var. *austrina*
    Syn = *Opuntia austrina*
    Syn = *Opuntia humifusa*
Opuntia compressa var. *macrorhiza*
    Syn = *Opuntia macrorhiza*
Opuntia cymochila
    Syn = *Opuntia mackensenii*
    Syn = *Opuntia tortispina*
    Syn = *Opuntia tortispina* var. *cymochila*
Opuntia davisii
    Syn = *Cylindropuntia davisii*
    Syn = *Opuntia tunicata* var. *davisii*
Opuntia drummondii
    Syn = *Cactus pusillus*
    Syn = *Opuntia pusilla*

Appendix

Opuntia emoryi
    Syn = ***Corynopuntia emoryi***
    Syn = *Grusonia emoryi*
Opuntia engelmannii
    Syn = ***Opuntia engelmannii* var. *alta***
    Syn = ***Opuntia engelmannii* var. *engelmannii***
Opuntia engelmannii var. alta
    Syn = *Opuntia lindheimeri*
Opuntia engelmannii var. engelmannii
    Syn = *Opuntia discata*
    Syn = *Opuntia engelmannii*
    Syn = *Opuntia engelmannii* var. *discata*
    Syn = *Opuntia lindheimeri*
    Syn = *Opuntia phaeacantha* var. *discata*
Opuntia engelmannii var. lindheimeri
    Syn = *Opuntia lindheimeri*
Opuntia engelmannii var. linguiformis
    Syn = *Opuntia lindheimeri* var. *linguiformis*
    Syn = *Opuntia linguiformis*
Opuntia grahamii
    Syn = ***Corynopuntia grahamii***
    Syn = *Grusonia grahamii*
Opuntia humifusa
    Syn = *Opuntia compressa* var. *humifusa*
Opuntia imbricata var. *argentea*
    Syn = ***Cylindropuntia imbricata* var. *argentea***
Opuntia imbricata var. *imbricata*
    Syn = *Cylindropuntia imbricata* var. *arborescens*
    Syn = ***Cylindropuntia imbricata* var. *imbricata***
Opuntia kleiniae
    Syn = ***Cylindropuntia kleiniae***
Opuntia leptocaulis
    Syn = ***Cylindropuntia leptocaulis***

Opuntia lindheimeri
    Syn = ***Opuntia engelmannii* var. *lindheimeri***
Opuntia lindheimeri var. *aciculata*
    Syn = ***Opuntia aciculata***
Opuntia lindheimeri var. *chisosensis*
    Syn = ***Opuntia chisosensis***
Opuntia lindheimeri var. *linguiformis*
    Syn = ***Opuntia engelmannii* var. *linguiformis***
Opuntia linguiformis
    Syn = ***Opuntia engelmannii* var. *linguiformis***
Opuntia mackensenii
    Syn = *Opuntia cymochila*
    Syn = ***Opuntia tortispina***
    Syn = *Opuntia tortispina* var. *cymochila*
Opuntia macrocentra
    Syn = *Opuntia violacea*
    Syn = *Opuntia violacea* var. *macrocentra*
Opuntia macrorhiza
    Syn = *Opuntia compressa* var. *macrorhiza*
Opuntia macrorhiza var. *pottsii*
    Syn = ***Opuntia pottsii***
    Syn = *Opuntia stenochila*
Opuntia microdasys var. *rufida*
    Syn = ***Opuntia rufida***
Opuntia palmeri
    Syn = ***Opuntia chlorotica***
Opuntia phaeacantha
    Syn = *Opuntia phaeacantha* var. *major*
    Syn = *Opuntia phaeacantha* var. *phaeacantha*
Opuntia phaeacantha var. *camanchica*
    Syn = ***Opuntia camanchica***
Opuntia phaeacantha var. *major*
    Syn = ***Opuntia phaeacantha***
    Syn = *Opuntia phaeacantha* var. *phaeacantha*

Opuntia phaeacantha var. *phaeacantha*
    Syn = *Opuntia phaeacantha*
    Syn = *Opuntia phaeacantha* var. *major*
Opuntia phaeacantha var. *spinosibacca*
    Syn = *Opuntia spinosibacca*
Opuntia polyacantha* var. *trichophora
    Syn = *Opuntia trichophora*
Opuntia pottsii
    Syn = *Opuntia macrorhiza* var. *pottsii*
    Syn = *Opuntia stenochila*
Opuntia pusilla
    Syn = *Cactus pusillus*
    Syn = *Opuntia drummondii*
Opuntia rufida
    Syn = *Opuntia microdasys* var. *rufida*
    Syn = *Opuntia rufida* var. *tortiflora*
Opuntia schottii
    Syn = *Corynopuntia schottii*
    Syn = *Grusonia schottii*
Opuntia spinosibacca
    Syn = *Opuntia phaeacantha* var. *spinosibacca*
Opuntia stenochila
    Syn = *Opuntia macrorhiza* var. *pottsii*
    Syn = *Opuntia pottsii*
Opuntia stricta
    Syn = *Cactus strictus*
    Syn = *Opuntia stricta* var. *dillenii*
Opuntia stricta var. *dillenii*
    Syn = *Cactus strictus*
    Syn = *Opuntia stricta*
Opuntia tortispina
    Syn = *Opuntia cymochila*
    Syn = *Opuntia mackensenii*
    Syn = *Opuntia tenuispina*
    Syn = *Opuntia tortispina* var. *cymochila*
Opuntia tortispina var. *cymochila*
    Syn = *Opuntia cymochila*
    Syn = *Opuntia mackensenii*
    Syn = *Opuntia tortispina*
Opuntia trichophora
    Syn = *Opuntia polyacantha* var. *trichophora*
Opuntia tunicata
    Syn = *Cylindropuntia tunicata*
Opuntia violacea
    Syn = *Opuntia azurea*
    Syn = *Opuntia azurea* var. *diplopurpurea*
    Syn = *Opuntia azurea* var. *parva*
    Syn = *Opuntia macrocentra*
    Syn = *Opuntia violacea* var. *macrocentra*
Opuntia violacea var. *macrocentra*
    Syn = *Opuntia azurea*
    Syn = *Opuntia azurea* var. *diplopurpurea*
    Syn = *Opuntia azurea* var. *parva*
    Syn = *Opuntia macrocentra*
    Syn = *Opuntia violacea*
Peniocereus greggii
    Syn = *Cereus greggii*
    Syn = *Cereus pottsii*
Sclerocactus brevihamatus var. tobuschii
    Syn = *Ancistrocactus tobuschii*
    Syn = *Mammillaria tobuschii*
Sclerocactus tobuschii
Sclerocactus intertextus var. dasyacanthus
    Syn = *Echinocactus intertextus* var. *dasyacanthus*
    Syn = *Echinomastus intertextus*
    Syn = *Echinomastus intertextus* var. *dasyacanthus*
Sclerocactus intertextus var. intertextus
    Syn = *Echinocactus intertextus*
    Syn = *Echinomastus intertextus*
    Syn = *Echinomastus intertextus* var. *intertextus*

Sclerocactus mariposensis
    Syn = *Echinocactus mariposensis*
    Syn = *Echinomastus mariposensis*
    Syn = *Neolloydia mariposensis*
Sclerocactus scheeri
    Syn = *Ancistrocactus scheeri*
    Syn = *Echinocactus scheeri*
Sclerocactus tobuschii
    Syn = *Ancistrocactus tobuschii*
    Syn = *Mammillaria tobuschii*
    Syn = *Sclerocactus brevihamatus* var. *tobuschii*
Sclerocactus uncinatus* var. *wrightii
    Syn = *Ancistrocactus uncinatus* var. *wrightii*
    Syn = *Echinocactus uncinatus* var. *wrightii*
    Syn = *Glandulicactus uncinatus* var. *wrightii*

Sclerocactus warnockii
    Syn = *Echinomastus warnockii*
    Syn = *Neolloydia warnockii*
Thelocactus bicolor* var. *bicolor
    Syn = *Ferocactus bicolor*
    Syn = *Thelocactus pottsii*
Thelocactus bicolor* var. *flavidispinus
    Syn = *Thelocactus flavidispinus*
Thelocactus flavidispinus
    Syn = *Thelocactus bicolor* var. *bicolor*
Thelocactus setispinus
    Syn = *Hamatocactus bicolor*
    Syn = *Hamatocactus setispinus*
Wilcoxia poselgeri
    Syn = *Cereus poselgeri*
    Syn = *Echinocereus poselgeri*

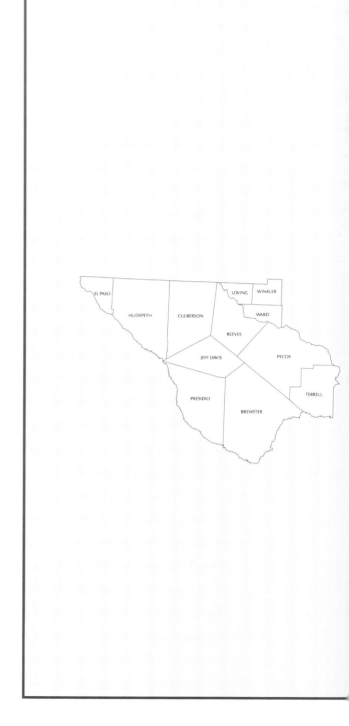

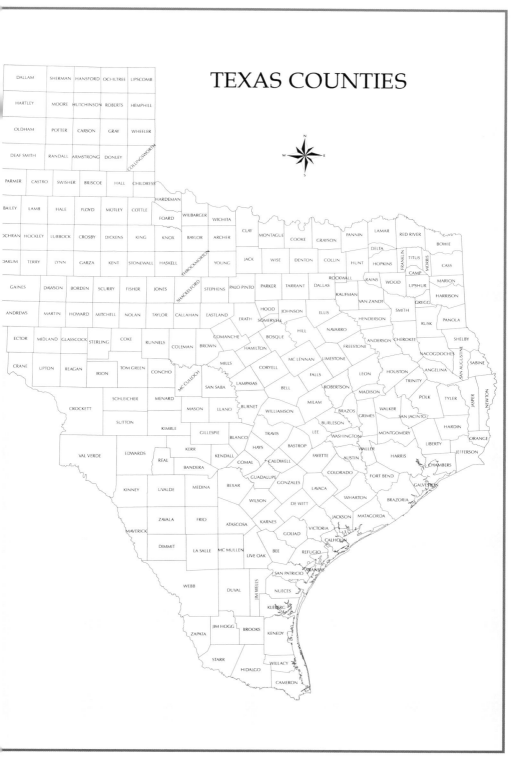

Bibliography

Anderson, Edward F. *The Cactus Family.* Portland, Ore.: Timber Press, 2001.

Anderson, Miles. *The World Encyclopedia of Cacti and Succulents.* New York: Hermes House, 1999.

Benson, Lyman. *Cactaceae.* In *Flora of Texas*, vol. 2., pt. 2, edited by C. L. Lundell et al., 221–317. Renner: Texas Research Foundation, 1969.

———. *The Cacti of the United States and Canada.* Stanford, Calif.: Stanford University Press, 1982.

———. *The Native Cacti of California.* Stanford, Calif.: Stanford University Press, 1969.

Bowers, Janice Emily. *Shrubs and Trees of the Southwest Deserts.* Tucson, Ariz.: Southwest Parks and Monuments Association, 1993.

Britton, N. L., and J. N. Rose. *The Cactaceae.* 4 vols. New York: Dover, 1963.

Carr, William R. "No Place Else but Texas." Unpublished draft. Austin: The Nature Conservancy of Texas, 2002.

Champie, Clark. *Cacti and Succulents of El Paso.* Santa Barbara, Calif.: Abbey Garden Press, 1974.

Damude, Noreen, and Kelly Conrad Bender. *Texas Wildscapes.* Austin: Texas Parks and Wildlife Press, 1999.

Diggs, George M., Jr., Barney L. Lipscomb, and Robert J. O'Kennon. *Shinners & Mahler's Illustrated Flora of North Central Texas.* Fort Worth: Botanical Research Institute of Texas, 1999.

Dodge, Natt N. *Flowers of the Southwest Deserts.* Tucson, Ariz.: Southwest Parks and Monuments Association, 1985.

Earle, W. Hubert. *Cacti of the Southwest.* Tempe, Ariz.: Rancho Arroyo Book Distributor, 1980.

Eggli, Urs. *Glossary of Botanical Terms with Special Reference to Succulent Plants.* Richmond, Surrey, U.K.: British Cactus & Succulent Society, 1993.

Ellis, John. *Beginning Guide to Cacti and Succulents.* New York: Sterling Publishing, 2004.

Evans, Douglas B. *Cactuses of Big Bend National Park.* Austin: University of Texas Press, 1998.

Everitt, James H., and D. Lynn Drawe. *Trees, Shrubs and Cacti of South Texas.* Lubbock: Texas Tech University Press, 1993.

Fernald, Merritt L. *Gray's Manual of Botany.* New York: American Book Co., 1950.

Fischer, Pierre C. *70 Common Cacti of the Southwest.* Tucson, Ariz.: Southwest Parks and Monuments Association, 1989.

Flora of North America Editorial Committee, eds. *Flora of North America North of Mexico.* Vol. 4. New York and Oxford: Oxford University Press, 2003.

Glass, Charles, Clive Innes, and Marcus Schneck. *Identifying Cacti.* Edison, N.J.: Chartwell Books, 1996.

Gould, Frank W. *The Grasses of Texas.* College Station: Texas A&M University Press, 1975.

Hatch, Stephan L., K. N. Gandhi, and Larry E. Brown. *Checklist of the Vascular Plants of Texas*. College Station: Texas A&M University Press, 1997.

Hatch, Stephan L., and Jennifer Pluhar. *Texas Range Plants*. College Station: Texas A&M University Press, 1999.

Hecht, Hans. *Cacti & Succulents*. New York: Sterling Publishing, 1994.

Hewitt, Terry. *The Complete Book of Cacti and Succulents*. New York: DK Publishing, 1997.

———. *101 Essential Tips—Cacti and Succulents*. New York: Dorling Kindersley Publishing, 1996.

Huey, George H. H., and Rose Houk. *Wild Cactus*. New York: Artisan, 1996.

Hunt, David. *The New Cactus Lexicon*. Milborne Port, Somerset, U.K.: dh Books, 2006.

Innes, Clive, and Charles Glass. *The Illustrated Encyclopedia of Cacti*. Edison, N.J.: Chartwell Books, 2001.

Jones, Stanley D., Joseph K. Wipff, and Paul M. Montgomery. *Vascular Plants of Texas*. Austin: University of Texas Press, 1997.

Loflin, Brian, and Shirley Loflin. *Grasses of the Texas Hill Country*. College Station: Texas A&M University Press, 2005.

Luebbermann, Mimi. *Cactus & Succulents*. San Francisco: Chronicle Books, 1997.

MacLeod, William. *Big Bend Vistas*. Alpine: Texas Geological Press, 2002.

McMahan, Craig A., Roy G. Frye, and Kirby L. Brown. *The Vegetation Types of Texas*. Austin: Texas Parks and Wildlife, 1984.

Maddox, Ed. *10,000 Miles Hunting Cactus in Texas*. Self-published, 1984.

Manke, Elizabeth. *Cactus, the Most Beautiful Varieties and How to Keep Them Healthy*. Hauppauge, N.Y.: Barrons, 2000.

Mauseth, James D. *Plant Anatomy*. Menlo Park, Calif.: Benjamin/Cummings Publishing, 1988.

———. "Structure-Function Relationships in Highly Modified Shoots of Cactaceae." *Annals of Botany* (2005) 98: 901–26.

Mauseth, James D., Roberto Kiesling, and Carlos Ostolaza. *A Cactus Odyssey*. Portland, Ore.: Timber Press, 2002.

Pizzetti, Mariella. *Simon and Schuster's Guide to Cacti and Succulents*. New York: Simon and Schuster, 1986.

Poole, Jackie M., William R. Carr, Dana M. Price, and Jason R. Singhurst. *Rare Plants of Texas: A Field Guide*. College Station: Texas A&M University Press, 2007.

Powell, A. Michael, and James F. Weedin. *Cacti of the Trans-Pecos & Adjacent Areas*. Lubbock: Texas Tech University Press, 2004.

Powell, A. Michael, James F. Weedin, and Shirley A. Powell. *Cacti of Texas: A Field Guide*. Lubbock: Texas Tech University Press, 2008.

Quinn, Meg. *Cacti of the Desert Southwest*. Tucson, Ariz.: Rio Nuevo Publishers / Treasure Chest Books, 2001.

Sajeva, Maurizio, and Mariangela Costanzo. *Succulents, the Illustrated Dictionary*. Portland, Ore.: Timber Press, 1994.

Schuster, Danny. *The World of Cacti*. New York: Facts on File, 1990.

Slaba, Rudolf. *The Illustrated Guide to Cacti*. New York: Sterling Publishing, 1992.

Turner, B. L., Holly Nichols, Geoffrey Denny, and Oded Doron. *Atlas of the Vascular Plants of Texas*. Fort Worth: Botanical Research Institute of Texas, 2003.

United States Department of Agriculture. S*oil Survey of Comal and Hays Counties, Texas*. Washington, D.C.: USDA, 1984.

Vines, Robert A. *Trees, Shrubs, and Woody Plants of the Southwest*. Austin: University of Texas Press, 1960.

Wasowski, Sally, and Andy Wasowski. *Native Texas Plants*. Houston: Gulf Publishing, 1991.

Wauer, Roland H. *A Naturalist's Mexico*. College Station: Texas A&M University Press, 1992.

———. *Naturalist's Big Bend*. College Station: Texas A&M University Press, 1980.

Weniger, Del. *Cacti of Texas and Neighboring States*. Austin: University of Texas Press, 1988.

———. *Cacti of the Southwest*. Austin: University of Texas Press, 1969.

Whitson, Tom D., Larry C. Burrill, Steven A. Dewey, David W. Cudney, B. E. Nelson, Richard D. Lee, and Robert Parker. *Weeds of the West*. Jackson: University of Wyoming, 1992.

Wrede, Jan. *Trees, Shrubs and Vines of the Texas Hill Country*. College Station: Texas A&M University Press, 2005.

Glossary

Acicular: Needle shaped, narrow, pointed.

Acidic: Sour or corrosive quality; opposite of alkaline or basic; frequently, as a property of soils.

Aerial root: Root that is above ground, growing from stems, taking nutrients from the air.

Alternate: Arranged singly at each node; set opposite on a stem or branch.

Angled: Not rounded, usually triangular throughout the long axis.

Annulate: Having circling rings or projecting bands that mark growth increments.

Anther: The top end of a stamen that contains the pollen emanating from pollen sacs.

Apex: The top of a structure.

Apical: Refers to the apex or the tip of a stem.

Appressed: Lying flat against the stem of the plant.

Arborescent: Treelike in shape, usually with a distinct trunk at the base of the plant.

Areolar gland: A nectar-secreting gland found on the surface of the areoles of some species; an extrafloral nectary.

Areole: The woolly, cushionlike structure in cacti from which spines and sharp bristles grow, as well as flowers, fruits, and roots; an axillary bud.

Armed: In cacti, refers to the growth of sharp spines and bristles for protection.

Articulate: Jointed parts, connected at a node.

Ascending: Rising upward.

Axil: The upper angle between a branch and the stem.

Basal: Near the point of origin or lowest part.

Basic: Alkaline, not neutral or acidic; frequently, referring to soils containing soluble mineral salts.

Berry: A many-seeded succulent fruit.

Bract: A modified leaf growing beneath a flower that may take different shapes, colors, and consistencies.

Bud: A flower that has not opened from which stems, leaves, flowers, seeds, or fruits may develop.

Bulbous: Swollen at the base.

Caballos novaculite: Variety of sedimentary chert found in uplifts of the Marathon Basin and again in the Ouachita Mountains of Arkansas.

Calcareous: Containing calcium carbonate, calcium, or lime as in limestone or its soil derivatives.

Calyx: The outer flower segments, which are its sepals, that protect the flower bud and its internal components.

Campanulate: Bell shaped.

Carpel: The structure that encloses the ovules in flowering plants, including the ovary, stigma, and style.

Central spine: A spine that is clearly different from other spines by being longer and thicker or even a different color; most frequently near the areole's center.

Cespitose: Formed into numerous stems into dense mats or clumps.

Cholla: Cacti that have upright or semi-upright growth with jointed and cylindrical stems.

Clavate: Shaped like a club with a narrow base and swelling toward the top.

Confluent: Running together; merging.

Corolla: Very conspicuous part of a flower within the calyx, the petals.

Cortex: The tissue region between the epidermis and the vascular tissue in a plant stem; quite thick and vascularized in cacti, and outermost region capable of photosynthesis.

Curved: Bent smoothly from the base to the tip.

Deciduous: Having a limited lifespan or persistence; most often refers to plants that drop their leaves before winter; in cacti, refers to organs or parts that drop for whatever reason.

Decumbent: Reclining, trailing, or lying down on the ground, often with a rising apical part or tip.

Deflexed: Sharply curved or bent back or down upon itself.

Dehiscent: A splitting open of the fruit, seed case, or pod when mature.

Descending: Angling downward.

Discoid: Disk shaped; having a rounded and flattened structure like some seeds.

Distal: Most distant from the point of attachment.

Divergent: Growing away from one another, especially from an areole.

Endemic: A species growing in an area with limited geographic distribution.

Ephemeral: A plant or structure with a short life cycle.

Epidermis: The outermost layer of cells on a plant meant to protect the tissue underneath; may include a waxy, nonliving layer on top for extra protection.

Felt: Thick covering of intertwining hairs or matted filaments.

Fibrous roots: Several major roots of the same size and characteristics growing from the same area.

Filament: The threadlike part of the stamen; slender stalk.

Flattened: Compressed in one direction along the long dimension.

Flower: The reproductive, seed-producing plant structure that attracts pollinating insects and other agents.

Funnelform: Shaped like a funnel.

Genus (pl., genera): A grouping of species having similar characteristics and unique enough to be distinct from other groups of species.

Glabrous: Not hairy; smooth.

Glaucous: Having a waxy whitish to grayish blue substance on the surface that can be rubbed off.

Globose: Rounded or spherical; globelike.

Glochid: A short, barbed bristle that grows from the areoles of opuntia cacti; occurring in tufts that detach very easily.

Hilum: A scar or depression on a seed where it was once attached while growing; the basal point of a seed.

Hooked: Apically recurved; curved backward like a fishhook.

Hybrid: A cross between two different species.

Joint: One stem segment; typical of opuntias, which frequently have several stem segments.

Keeled: Having a projecting ridge on a structure as in the keel of a boat.

Lateral: Positioned on the side or extending to the side.

Glossary

Latex: A milky fluid that oozes from plant tissues when cut or damaged.

Leaf: An outgrowth or appendage attached to a stem or branch; carries out photosynthesis, but which is accomplished in most cacti by the stems and branches.

Meristem: Type of tissue that has the ability to divide and to differentiate.

Node: The knotlike stem location commonly bearing branches or leaves.

Obovate: Egg shaped, with the widest part above the middle.

Ovary: The expanded part of the pistil that contains the ovules and that will develop into a fruit.

Ovate: Egg shaped with the widest part near the base.

Ovule: Immature seed not yet fertilized and contained within the ovary.

Papillate: Having minute raised, rounded projections.

Pectinate: Shaped like the teeth on a comb.

Petal: A structure of the corolla; modified leaf; several can radiate around the center of a flower; often colorful and showy.

Pistil: The female seed-bearing structure of a flower.

Pitted: Having minute, rounded depressions.

Plumose: Featherlike.

Pollen: Tiny spore granules that emanate from the anthers of the stamen, and, typically male, fertilize the ovules.

Pollen tube: A slender conduit in the style that conducts pollen from the stigma to the ovary.

Porrect: Perpendicular to the surface of the plant.

Prostrate: Lying on, or close to, the ground.

Proximal: Closest to the point of origin.

Pubescent: Covered with short, fine, velvety hairs.

Pyriform: Pear shaped.

Radial: At the edge of a circular or oval area.

Radial spine: One of a group of spines arranged at or near the margins of an areole, generally surrounding the central spines when present.

Radiating: Positioned outward like rays or spokes.

Reclining: Lying, sprawling, or leaning.

Recurved: Curving downward, backward, or back against itself.

Reflexed: Turned abruptly backward; bent or folding back.

Reniform: Kidney shaped.

Rib: A raised surface that runs vertically or spirals downward; arises from stem tissue; composed of stem tissue and tubercles.

Ridged: Ringed or corrugated across the long axis.

Saline: Salty; containing dissolved salts.

Savannah: Grassland with scattered trees and shrubs in clumps or in thickets.

Scabrous: Rough to the touch.

Selaginella: Genus of spikemoss plants found in the caballos novaculite of the Marathon Basin; microhabitat of cacti unique to the area.

Sepal: In cacti, one of the nonshowy outer petals of the flower; part of the corolla; the outer circle of the flower.

Setaceous: Set with bristles; bristlelike.

Shoot: Above ground stem.

Spatulate: Spoon shaped or spatula shaped.

Species: A genetically closely related population having recognizable characteristics and form.

Spine: In cacti, the hard, sharp outgrowth; may be rigid, woody, flexible, or hairlike; grows from the areole and considered a modified leaf.

Stamen: The male, pollen-providing organ of the flower.

Stigma: The tip of the pistil supported by the style; structure that receives the pollen; frequently multilobed in cacti.

Stoma (pl., stomata): An opening, or pore, from which respiration and transpiration take place.

Striated: Marked with longitudinal lines, grooves, or ridges.

Style: Slender tissue column supporting the stigma, arising from the ovary, and containing the pollen tube.

Succulent: A plant that stores water in fleshy tissue of leaves or stems; usually drought resistant.

Taproot: Primary root growing deep below soil level that provides nutrients to the plant and water storage.

Taxon (pl., taxa): A taxonomic group of plants of any rank.

Terete: Circular throughout the long axis.

Terminal: Growing at or positioned at the distal extremity; at the tip or end.

Transpiration: Giving off water vapor by way of the stomata of a plant.

Trichome: A hairlike outgrowth from the plant's own epidermis; usually multicellular.

Tuber: Thickened, underground stem that may enlarge to store water and nutrients.

Tubercle: A raised protuberance on the plant stem, usually supporting an areole.

Turgid: Swollen or distended by internal pressure.

Twisted: Rotated about the long axis.

Umbilicate apex: Depressed and scarred center where the flower was attached to the top of the fruit.

Umbilicus: A cuplike depression at the tip end of a fruit.

Variety: A taxonomic subdivision of a species.

Vascular tissue: Plant tissue that transports water and sugars.

Vascularized: Containing vascular tissue.

Verticil: A whorl.

Villous: Having interlaced hairs that appear soft, long, and closely spaced.

Wool: Intertwined or matted hairs.

Xeric: Extremely dry and arid.

Xerophyte: A plant with the ability to live and thrive in dry and arid conditions.

Index

Boldface type represents currently accepted cactus names used within the text and their respective pages.

abrojo, 61
Acanthocereus, 43
    pentagonus, 121
    ***tetragonus,* 121**
alicoche, 123, 145
alkaloid, hallucinogenic, 209
Ancistrocactus brevihamatus, 255
    scheeri, 193
    tobuschii, 257
    uncinatus var. *wrightii,* 195
Anderson, Edward, xiii, 259
ants, 29
apical growth, 20
apparently secure conservation status, 38
areoles, 21
Ariocarpus, 43, 44
    ***fissuratus,* 205**
Arizona barrel cactus, 219
ashy white pitaya, 149
Astrophytum, 43, 44
    ***asterias,* 207**

barbwire cactus, 121
bark, 18
barrel cactus, 43
bearded prickly pear, 117
bee, 30, 31
bee assassin, 29
beehive cactus, 133
Benson, Lyman, xiii
Berlandier, Jean Louis, xiii
Berlandier's hedgehog cactus, 127
berry, 26
bicolor cactus, 199
Big Bend cane cholla, 65

Big Bend cholla, 65
Big Bend country, 14
Big Bend eggs, 169
Big hill prickly pear, 79
big nipple cactus, 171
big nipple cory cactus, 131
big nipple, 133
big-needle pincushion cactus, 131
bird foot cactus, 173
biscuit cactus, 167
bisnaga de dulce, 211
black-and-yellow-spined prickly pear, 73
Blackland Prairies, 5
blind pear, 111
blind prickly pear, 111
blue barrel cactus, 211
Boke's button cactus, 231
Boquillas button cactus, 231
Boquillas Canyon, 75
Britton, Nathaniel, xiii
brown flowered hedgehog, 195
brown-spined prickly pear, 103
bud, cactus, 27
bunched cory cactus, 225
Burbank, Luther, 34
butterflies, 31
button cactus, 43, 233

caballos novaculite, 229
Cactaceae, 15
cactus, 15
    Arizona barrel, 219
    barbwire, 121
    barrel, 43
    beehive, 133
    Berlandier's hedgehog, 127
    bicolor, 199
    big nipple cory, 131
    big nipple, 171
    big-needle pincushion, 131

cactus (cont.)
   bird foot, 173
   biscuit, 167
   blue barrel, 211
   Boke's button, 231
   Boquillas button, 231
   bunched cory, 225
   button, 233
   button, 43
   carpet foxtail, 177
   cat claw, 195
   Chaffey's pincushion, 167
   chaparral, 125
   Chisos hedgehog, 137
   Chisos Mountain hedgehog, 137
   Christmas, 69
   claret cup, 139
   cob cory, 179
   cob, 179
   cocklebur, 109
   cock-spur, 109
   common button, 233
   cone, 185
   Correll hedgehog, 159
   Correll's green-flowered hedgehog, 159
   cory, 133
   dahlia, 129
   Davis hedgehog, 157, 165
   Davis' cholla, 61
   desert pincushion, 169
   devil's cholla, 55
   discus, 91
   divine, 209
   dog, 59
   Duncan's pincushion, 235
   dwarf cory, 173
   dwarf hedgehog, 229
   eagle-claw, 211
   eastern beehive, 241
   Fendler's hedgehog, 147
   finger, 227
   fishhook, 203
   foxtail, 183
   fragrant, 181

   giant fishhook, 215
   golden rainbow hedgehog, 141
   Graham dog, 57
   Graham's fishhook, 243
   grape, 251
   green-flowered hedgehog, 155
   greenflowered torch, 157
   Guadalupe pincushion, 177
   hair-covered, 251
   hair-grooved, 251
   hedgehog cory, 221
   hedgehog, 43, 203
   Hester's foxtail, 237
   Hester's pincushion, 237
   Heyder's pincushion, 245
   horse-crippler, 213
   junior Tom Thumb, 171
   lace hedgehog, 151
   lace spine, 249
   lady finger, 123
   Langtry claret cup, 139
   Langtry rainbow, 149
   Lee's pincushion, 177
   living rock, 43, **205**
   Lloyd's hedgehog, 165
   Lloyd's mariposa, 191
   long mamma nipple, 253
   longcentral woven-spine pineapple, 187
   Lower Rio Grande Valley barrel, 217
   Mariposa, 191
   Mexican strawberry, 143
   Missouri foxtail, 239
   multi-stemmed sea urchin, 223
   Nellie's pincushion, 173
   New Mexico beehive cactus, 181
   New Mexico rainbow, 171
   nipple beehive, 131
   nipple, 227, 247
   nylon hedgehog, 155
   pancake, 245
   pencil, 69
   pincushion, 43, 241

Index 283

cactus *(cont.)*
   pincushion-beehive, 43
   pineapple, 135, 227
   Pott's nipple, 183
   purple hedgehog, 147
   rainbow, 141
   rat-tail, 183
   rhinoceros, 221
   root, 193
   sandbur, 109
   scarlet hedgehog, 139
   Scheer's fishhook, 193
   sea urchin, 207, 221
   Senita, 32
   Silverlace, 175, 177
   small-flowered hedgehog, 163
   Sneed's pincushion, 177
   snipe, 255
   snowcone nipple, 175
   Solitario green-flowered hedgehog, 157
   sour, 181
   spiny hedgehog, 141
   spiny-star, 181
   star, 43, 205, 207
   starvation, 105
   strawberry hedgehog, 145
   strawberry pitaya, 143
   strawberry, 139, 153
   Texas claret cup, 139
   Texas cone, 185
   Texas rainbow, 141
   Texas, 185
   Tobusch fishhook, 257
   tree, 63
   triangle, 43, 121
   turk's head, 211
   twisted rib, 203
   varicolor cob, 179
   Warnock's, **197**
   western green-flowered hedgehog, 159
   whisker brush pincushion, 225
   white column foxtail, 179
   white flowered, 197
   whitespine cob, 175
   woven-spine pineapple, 189
cactus anatomy, 15
Cactus inermis, 115
cactus moth, 31
Cactus pusillus, 109
   strictus, 115
cactus wren, 33
Camanchican prickly pear, 83
candle cholla, 67
cane cholla, 63
Caprock Escarpment, 11
Caribbean Tropical Forest, 4
carpet foxtail cactus, 177
cat claw cactus, 195
Cereus berlandieri, 127
   fendleri, 147
   greggii, 125
   pentagonus, 121
   pentalophus, 123
   poselgeri, 129
   pottsii, 125
   stramineus, 153
Chaffey's pincushion cactus, 167
chain prickly pear, 101
chaparral cactus, 125
chaparral, Southwestern, 14
Chihuahuan desert scrub, 14
Chihuahuan Desert, 14
Chisos hedgehog cactus, 137
Chisos mountain hedgehog cactus, 137
Chisos pitaya, 137
Chisos prickly pear, 85
cholla, 43, 49
   Big Bend, 65
   candle, 67
   cane, 63
   Christmas, 69
   clumped dog, 53
   common devil, 55
   creeping, 55
   Davis,' 61
   devil's, 55
   dog, 43
   Graham's club, 57

cholla (*cont.*)
    icicle, 71
    Klein, 67
    Klein's pencil, 67
    mounded dwarf, 57
    Schott's dog, 59
    Schott's dwarf, 59
    sheathed, 71
    silver-spine cane, 65
    Stanley's, 55
    thistle, 61, 71
    tree, 63
    walking stick, 63
Christmas cactus, 69
Christmas cholla, 69
cinder bells, 159
cinnamon pear, 111
claret cup cactus, 139
clavellina, 59, 71
clockface prickly pear, 87
clumped dog cholla, 53
cob cactus, 175, 179
cob cory cactus, 179
cochineal bugs, 30
cocklebur cactus, 109
cock-spur cactus, 109
Comanche prickly pear, 83
comb hedgehog, 149
common button cactus, 233
common devil cholla, 55
compass barrel, 219
cone cactus, 185, 43
conservation status, 37
cork cambium, 18
corn cob escobaria, 179
Correll hedgehog cactus, 161
Correll's green-flowered hedgehog cactus, 161
cortex, 18
cortical bundles, 18, 19
cory cactus, 133, 225
Corynopuntia, 43, 44
    aggeria, **53**
    emoryi, **55**
    grahamii, **57**
    schottii, **59**

Coryphantha, 43, 44
    albicolumnaria, 175
    chaffeyi, 167
    dasyacantha, 169
    duncanii, 235
    ***echinus* var. *robusta*, 223**
    ***echinus*, 221**
    hesteri, 237
    ***macromeris* var. *runyonii*, 133**
    macromeris, 131, 133
    minima, 173
    missouriensis, 239
    neomexicana, 181
    pectinata, 221
    pottsiana, 171
    pottsii, 183
    radiosa var. *neomexicana*, 181
    ***ramillosa*, 225**
    ***robustispina* var. *scheeri*, 135**
    scheeri, 135
    sneedii var. *albicolumnaria*, 175
    sneedii var. *sneedii*, 177
    ***sulcata*, 227**
    tuberculosa, 179
    vivipara, 241
cow-tongue prickly pear, 95
crack star, 205
creeping cholla, 55
creosotebush, 14
critically imperiled conservation status, 38
Cross Timbers and Prairies, 6
crow-foot prickly pear, 109
Cylindropuntia, 43, 45
    davisii, **61**
    imbricata var. *arborescens*, **63**
    imbricata var. *argentea*, **65**
    kleiniae, **67**
    leptocaulis, **69**
    tunicata, **71**

dahlia cactus, 129
dark spined opuntia, 73

Index

Davis hedgehog cactus, 157, 165, 229
Davis' cholla, 61
dense mammillaria, 169
desert grasslands, 10
desert night-blooming cereus, 125
desert pincushion cactus, 169
devil's cholla, 55
devil's head, 213
discus cactus, 91
divine cactus, 209
dog cactus, 59
dog cholla, 43
dry whiskey, 205, 209
Duncan's pincushion cactus, 235
Duncan's snowball, 235
dwarf cory cactus, 173
dwarf hedgehog cactus, 229

eagle-claw cactus, 195, 211
early bloomer, 187, 189
eastern beehive cactus, 241
eastern prickly pear, 97
Echinocactus, 43, 45
    asterias, 207
    conoideus, 185
    emoryi, 219
    ***horizonthalonius*, 211**
    intertextus var. *dasyacanthus*, 187
    intertextus, 189
    macromeris, 131
    mariposensis, 191
    pectinatus, 149
    pottsiana, 171
    reichenbachii, 151
    scheeri var. *brevihamatus*, 255
    scheeri, 193
    sinuatus, 217
    ***texensis*, 213**
    uncinatus var. *wrightii*, 195
    williamsii, 209
    wislizeni, 219
Echinocereus, 43, 46
    ***berlandieri*, 127**
    chaffeyi, 167
    ***chisosensis*, 137**

    chloranthus var. *chloranthus*, 159
    chloranthus var. *cylindricus*, 155, 163
    chloranthus, 159
    coccineus var. *aggregatus*, 139
    ***coccineus*, 139**
    dasyacantha var. *duncanii*, 235
    dasyacantha, 169
    ***dasyacanthus*, 141**
    davisii, 229
    dubius, 145
    enneacanthus forma *brevispinus*, 143
    ***enneacanthus* var. *brevispinus*, 143**
    ***enneacanthus* var. *enneacanthus*, 145**
    enneacanthus var. *stramineus*, 153
    fendleri var. *rectispinus*, 147
    ***fendleri*, 147**
    lloydii, 165
    pectinatus var. *dasyacanthus*, 141
    ***pectinatus* var. *wenigeri*, 149**
    ***pentalophus*, 123**
    ***poselgeri*, 129**
    reichenbachii var. *chisosensis*, 137
    ***reichenbachii*, 151**
    runyonii, 167
    stanleyi, 163
    ***stramineus* var. *stramineus*, 153**
    triglochidiatus var. *paucispinus*, 139
    ***viridiflorus*, 155**
    viridiflorus subsp. *correllii*, 161
    ***viridiflorus* var. *canus*, 157**
    ***viridiflorus* var. *chloranthus*, 159**
    ***viridiflorus* var. *correllii*, 161**
    ***viridiflorus* var. *cylindricus*, 163**
    ***viridiflorus* var. *davisii*, 229**

Echinocereus (cont.)
    viridiflorus var. *nova,* 157
    x roetteri var. neomexicana, 165
    x roetteri, 165
Echinomastus
    intertextus var. *dasyacanthus,* 187
    intertextus, 187, 189
    intertextus var. *intertextus,* 189
    mariposensis, 191
    warnockii, 197
Edwards Plateau, 8
Endangered Species Act, 37
Engelmann, George, xiii
Engelmann's prickly pear, 91
ephemeral leaves, 15
epidermis, 17
Epithelantha, 43, 46
    bokei, 231
    micromeris var. *bokei,* 231
    ***micromeris,** 231, 233*
Escobar, Numa, 47
Escobar, Romulo, 47
Escobaria, 43, 47
    ***dasyacantha* var. *chaffeyi,* 167**
    ***dasyacantha* var. *dasyacantha,* 169**
    ***duncanii,* 235**
    ***emskoetterana,* 171**
    ***hesteri,* 237**
    ***minima,* 173**
    ***missouriensis,* 239**
    robertii, 171
    runyonii, 171
    ***sneedii* var. *orcuttii,* 175**
    ***sneedii* var. *sneedii,* 177**
    ***tuberculosa,* 179**
    vivipara var. *neomexicana,* 181
    vivipara var. *vivipara,* 241
extrafloral nectaries, 23, 22, 29, 193
Fendler's hedgehog cactus, 147

Fendler's pitaya, 147
Ferocactus, 43, 47
    bicolor, 199
    ***hamatacanthus* var. *sinuatus,* 217**
    hamatacanthus var. *crasispinus,* 215
    ***hamatacanthus* var. *hamatacanthus,* 215**
    ***wislizeni,* 219**
finger cactus, 227
fishhook cactus, 43, 203, 243
flapjack prickly pear, 87
flat-spined thelocactus, 201
Flora of North America Editorial Committee, xiv
floristic associations, 1
flowers, 25
foxtail cactus, 183
fragrant cactus, 181
fruit, 26, 27
fuzzy mammillaria, 249

giant fishhook cactus, 215
Glandulicactus uncinatus var. *wrightii,* 195
glochids, 23
glory of Texas, 43, 199
golden rainbow hedgehog cactus, 141
golden-spined prickly pear, 75
golf ball cactus, 249
Graham dog cactus, 57
Graham's club cholla, 57
Graham's fishhook cactus, 243
grape cactus, 251
grassland prickly pear, 101
Great Plains Grasslands, 10
green-flowered hedgehog cactus, 161
green-flowered pitaya, 159
greenflowered pitaya, 163
greenflowered torch cactus, 159
Gregg, Josiah, xiii
grooved nipple cactus, 227

Index

Grusonia, 44
    aggeria, 53
    emoryi, 55
    grahamii, 57
    schottii, 59
Guadalupe Peak, 12
Guadalupe pincushion cactus, 177
Gulf Prairies and Marshes, 3

habitat, 1
hair-covered cactus, 251
hair-grooved cactus, 251
hallucinogenic alkaloid, 209
Hamatocactus bicolor, 203
    setispinus, 203
hedgehog cactus, 43, 203
hedgehog cory cactus, 221
Hester's foxtail cactus, 237
Hester's pincushion cactus, 237
Heyder's pincushion cactus, 245
High Plains, 11
hilum, 42
Homalocephala texensis, 213
honey bees, 31
horse-crippler cactus, 213
Hunt, David, xiv, 259

icicle cholla, 71
imperiled conservation status, 38
International Cactaceae Systematics Group, xiv

javelina, 34
jumping spiders, 32
junior Tom Thumb cactus, 171

Klein cholla, 67
Klein's pencil cholla, 67

lace cactus, 151
lace hedgehog cactus, 151
lace spine cactus, 249
lady finger cactus, 123
Langtry claret cup cactus, 139
Langtry rainbow cactus, 149
leaffooted bug, 30

leaves, 15, 24
lechuguilla, 14
Lee's pincushion cactus, 177
lemonvine, 43
Lindheimer, Ferdinand, xiii
Lindheimer's prickly pear, 93
lingua de vaca, 95
little chilies, 247
living rock cactus, 205
living rock cactus, 43, 205
living rock, 205, 43
lizard catcher, 243
Llano Estacado Escarpment, 11
Lloyd, Francis E., 48
Lloyd's hedgehog cactus, 165
Lloyd's mariposa cactus, 191
long mamma nipple cactus, 253
long mamma, 131, 133
longcentral woven-spine pineapple cactus, 187
long-spine prickly pear, 77, 81
long-spined purplish prickly pear, 77, 81, 99
Lophophora, 43, 47
    williamsii, **209**
low prickly pear, 97
Lower Rio Grande Valley barrel cactus, 217

Mammillaria, 43, 48
    albicolumnaria, 175
    conoidia, 185
    dasyacantha, 169
    denudata, 249
    duncanii, 235
    fissurata, 205
    grahamii var. *grahamii,* **243**
    hesteri, 237
    heyderi var. *hemispherica,* 245
    ***heyderi* var. *heyderi,* 245**
    ***heyderi* var. *meiacantha,* 247**
    heyderi, 245
    ***lasiacantha,* 249**
    longimamma var. *sphaerica,* 253
    macromeris, 131, 133

Mammillaria *(cont.)*
    meiacantha, 247
    microcarpa, 243
    micromeris, 233
    missouriensis, 239
    multiceps, 251
    nellieae, 173
    pectinata, 221
    ***pottsii,* 183**
    ***prolifera* var. *texana,* 251**
    pusilla var. *texana,* 251
    ramillosa, 225
    robustispina, 135
    runyonii, 247
    sneedii, 177
    ***sphaerica,* 253**
    sulcata, 227
    tobuschii, 257
    tuberculosa, 179
    vivipara var. *neomexicana,* 181
    vivipara, 241
manca caballo, 213
Marathon basin thelocactus, 201
marble fruit prickly pear, 117
Mariposa cactus, 191
Mariscal Mountain, 75
meristem, 15, 20
mescal button, 209
Mexican strawberry cactus, 143
Missouri foxtail cactus, 239
moths, 31
mound pitaya, 143
mounded dwarf cholla, 57
mound-forming opuntia, 53
mulato, 233
multi-stemmed sea urchin cactus, 223

Narnia, 30
needle "mulee" beehive, 135
Nellie's cory, 173
Nellie's pincushion cactus, 173
***Neolloydia,* 43, 48**
    ***conoidea,* 185**
    mariposensis, 191
    texensis, 185

warnockii, 197
New Cactus Lexicon, xiv, 259
New Mexico beehive cactus, 181
New Mexico prickly pear, 103
New Mexico rainbow cactus, 163
night-blooming cereus, 43, 121
nipple beehive cactus, 131
nipple cactus, 227, 245, 247
nipple cactus, 247
nipple, 133
nopal cegador, 111
nopal, 89, 91, 93
novaculite, 237
nylon hedgehog cactus, 155, 163

***Opuntia,* 43, 48**
    aggeria, 53
    ***atrispina,* 73**
    ***aureispina,* 75**
    azurea var. *aureispina,* 75
    ***azurea* var. *diplopurpurea,* 77**
    ***azurea* var. *discolor,* 79**
    ***azurea* var. *parva,* 81**
    azurea, 77, 81
    ***camanchica,* 83**
    ***chisosensis,* 85**
    ***chlorotica,* 87**
    compressa var. *humifusa* , 97
    compressa var. *macrorhiza,* 101
    cymochila, 119
    davisii, 61
    discata, 91
    drummondii, 109
    emoryi, 55
    ***engelmannii* var. *alta,* 89**
    engelmannii var. *discata,* 91
    ***engelmannii* var. *engel mannii,* 91**
    ***engelmannii* var. *lind heimeri* , 93**
    ***engelmannii* var. *lingui formis,* 95**
    engelmannii, 91
    grahamii, 57
    ***humifusa,* 97**
    imbricata var. *arborescens,* 63

Opuntia (cont.)
   imbricata var. *argentea,* 65
   kleiniae, 67
   leptocaulis, 69
   lindheimeri var. *chisosensis,* 85
   lindheimeri var. *linguiformis,* 95
   lindheimeri, 89, 91, 93
   linguiformis, 95
   mackensenii, 119
   macrocentra var. *aureispina,* 75
   ***macrocentra,* 99**
   macrorhiza var. *pottsii,* 107
   ***macrorhiza,* 101**
   microdasys var. *rufida,* 111
   palmeri, 87
   phaeacantha var. *camanchica,* 83
   phaeacantha var. *discata,* 91
   phaeacantha var. *major,* 83, 103
   phaeacantha var. *phaeacantha,* 103
   phaeacantha var. *spinosibacca,* 113
   ***phaeacantha,* 103**
   ***polyacantha* var. *trichophora,* 105**
   ***pottsii,* 107**
   ***pusilla,* 109**
   rufida var. *tortiflora,* 111
   ***rufida,* 111**
   schottii, 59
   ***spinosibacca,* 113**
   stapeliae, 71
   stenochila, 107
   stricta var.*dillenii,* 115
   ***stricta,* 115**
   ***strigil,* 117**
   tenuispina, 119
   tortispina var. *cymochila,* 119
   ***tortispina,* 119**
   trichophora, 105
   tunicata var. *davisii,* 61
   tunicata, 71
   violacea var. *macrocentra,* 77, 81, 99
   violacea, 99
   wrightii, 67

pale mammillaria, 253
pancake cactus, 245
pancake prickly pear, 87
Pecos River, 12
Peiresc, Nicholas de, 49
pencil cactus, 69, 129
***Peniocereus,* 43, 49**
   greggii, **125**
Pereskia, 43, 49
pest prickly pear, 115
peyote, 43, 209
pincushion cactus, 43, 241
pincushion-beehive cactus, 43
pineapple cactus, 135, 227
Pineywoods, 2
pitahaya, 139
pitaya, 143, 145
plains prickly pear, 101, 105, 119
pollen, 25
Post Oak Savannah, 4
Pott's mammillaria, 183
Pott's nipple cactus, 183
Pott's prickly pear, 107
Powell, A. Michael, xiv, 259
Prickly pear, 43, 49
   bearded, 117
   Big Hill, 79
   black-and-yellow-spined, 73
   blind, 111
   brown-spined, 103
   camanchican, 83
   chain, 101
   chisos, 85
   clockface, 87
   comanche, 83
   cow-tongue, 95
   crow-foot, 109
   eastern, 97
   flapjack, 87
   golden-spined, 75
   Lindheimer's, 93
   long-spine, 77, 81
   long-spined purplish, 77, 81, 99
   low, 97
   marble fruit, 117

Prickly pear *(cont.)*
    New Mexico, 103
    pancake, 87
    pest, 115
    plains, 101, 105, 119
    Pott's, 107
    purple, 77, 81
    purple-fruited, 91, 103
    red-spined, 113
    sandbur, 109
    smooth, 97
    southern plains, 105
    spiny-fruited, 113
    tall, 89
    Texas, 93
    tulip, 91, 103
    twisted spine plains, 119
    western, 101
    purple candle, 151
purple hedgehog cactus, 147
purple prickly pear, 77, 81
purple-fruited prickly pear, 91, 103

queen of the night, 43, 125

rainbow cactus, 141
rat-tail cactus, 183
rattlesnake, 33
red-spined prickly pear, 113
rhinoceros cactus, 221
ribs, 20
ridge, 42
Rio Grande Valley, 7, 8
Rocky Mountains, 13
Rolling Plains, 10
root cactus, 193
roots, 16
Rose, Joseph, xiii
Runyon's coryphantha, 133
Runyon's escobaria, 171

sacasil, 129
sand dollar, 207
sandbur cactus, 109
sandbur prickly pear, 109
scarlet hedgehog cactus, 139

Scheer's fishhook cactus, 193
Schott's dog cholla, 59
Schott's dwarf cholla, 59
Sclerocactus, 43, 50
    brevihamatus **var.** *tobuschii,*
    257
    brevihamatus, **255**
    intertextus **var.** *dasyacan-*
    thus, **187**
    intertextus **var.** *intertextus,*
    189
    mariposensis, **191**
    scheeri, **193**
    uncinatus **var.** *wrightii,* **195**
    warnockii, **197**
scorpion, 32
sea urchin cactus, 207, 221
secretory glands, 23
secure conservation status, 38
seeds, 28, 41
seed, cereus type, 28, 41
seed,, type, 28, 41
Selaginella, 229, 237
Selenicereus, 43, 50
Senita cactus, 32
Senita flower, 32
Senita moth, 32
sheathed cholla, 71
short-spined fishhook cactus, 255
silver column cactus, 191
silverlace cactus, 175, 177
silver-spine cane cholla, 65
small-flowered hedgehog cactus, 163
small-spined pincushion, 247
smooth prickly pear, 97
Sneed's pincushion cactus, 177
snipe cactus, 255
snowcone nipple cactus, 175
Solitario green-flowered hedgehog cactus, 157
Sotol, 14
sour cactus, 181
South Texas Brush Country, 7
South Texas Plains, 7
southern plains prickly pear, 105

southwestern barrel, 219
southwestern chaparral, 14
spider, jumping, 32
spines, 21, 39
spiny hedgehog cactus, 141
spiny-fruited prickly pear, 113
spiny-star cactus, 181
Stanley's cholla, 55
star cactus, 43, 205, **207**
star peyote, 207
starvation cactus, 105
stems, 17, 19, 38
stems, cross section, 19
Stockton Plateau, 9
stomata, 18
strawberry cactus, 139, 145, 147, **153**
strawberry hedgehog cactus, 145
strawberry hedgehog, 153
strawberry pitaya cactus, 143
subtropical woodlands, 7

tall prickly pear, 89
Tamaulipan Biotic Province, 7
Tamaulipan thorn scrub, 7
tapon, 233
tarantulas, 32
tasajillo, 69
testa, 42
Texas cactus, 185
Texas claret cup cactus, 139
Texas cone cactus, 185
Texas Hill Country, 9
Texas Panhandle, 10
Texas prickly pear, 93
Texas pride, 199
Texas rainbow cactus, 141
Texas Trans-Pecos, xiv
Thelocactus, 43, 50
    bicolor var. ***bicolor*, 199**
    bicolor var. ***flavidispinus*, 201**
    flavidispinus, 201
    pottsii, 199
    setispinus. 203

thistle cholla, 61, 71
Tobusch fishhook cactus, 257
Trans-Pecos, 12
tree cactus, 63
tree cholla, 63
triangle cactus, 121
triangle cactus, 43
tubercles, 20
tulip prickly pear, 91, 103
tuna, 26, 49, 93, 89, 93
turk's head cactus, 195, 211, 215
twisted rib cactus, 203
twisted spine plains prickly pear, 119

varicolor cob cactus, 179
Vegetational Areas of Texas, 1
visnaga, 211, 219
vulnerable conservation status, 38

walking stick cholla, 63
Warnock biznagita, 197
Warnock's cactus, 197
Weedin, James, xiv, 259
Weniger hedgehog, 149
western diamondback rattlesnake, 33
western green-flowered hedgehog cactus, 159
western prickly pear, 101
whisker brush pincushion cactus, 225
whiskered barrel, 215
white biznagita, 187, 189
white column foxtail cactus, 179
white column, 175
white flowered cactus, 197
white-flowered bisnagita, 189
white-flowered visnagita, 187
whitespine cob cactus, 175
Wilcoxia poselgeri, 129
Wislizenus, A., xiii
woven-spine pineapple cactus, 189